Common Conditions in Gynaecology

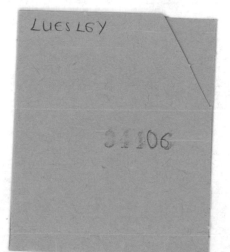

Common Conditions in Gynaecology

A problem-solving approach

Edited by David M. Luesley
Professor of Gynaecological Oncology
City Hospital NHS Trust
Birmingham, UK

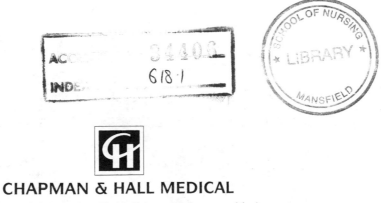

CHAPMAN & HALL MEDICAL

London · Weinheim · New York · Tokyo · Melbourne · Madras

Published by Chapman & Hall, 2–6 Boundary Row, London SE1 8HN, UK

Chapman & Hall, 2–6 Boundary Row, London SE1 8HN, UK

Chapman & Hall GmbH, Pappelallee 3, 69469 Weinheim, Germany

Chapman & Hall USA, 115 Fifth Avenue, New York, NY 10003, USA

Chapman & Hall Japan, ITP-Japan, Kyowa Building, 3F, 2-2-1 Hirakawacho, Chiyoda-ku, Tokyo 102, Japan

Chapman & Hall Australia, 102 Dodds Street, South Melbourne, Victoria 3205, Australia

Chapman & Hall India, R. Seshadri, 32 Second Main Road, CIT East, Madras 600 035, India

First edition 1997

© 1997 Chapman & Hall

© Anona L. Blackwell figures 6.1, 6.2 and 6.3

Typeset in 12/14 pt Garamond 3 by Best-set Typesetter Ltd., Hong Kong

Printed in Italy by Vincenzo Bona

ISBN   0 412 72380 8

A catalogue record for this book is available from the British Library

Library of Congress Catalog Card Number: 97-66879

Contents

Contributors

Mr M. Afnan
Senior Lecturer in Obstetrics and Gynaecology
Birmingham Women's Hospital
Queen Elizabeth Medical Centre
Edgbaston
Birmingham B15 2TG

Dr S. Blunt
Consultant Obstetrician and Gynaecologist
Birmingham Women's Hospital
Queen Elizabeth Medical Centre
Edgbaston
Birmingham B15 2TG

Dr C. Bradbeer
Consultant in Genitourinary Medicine
37 Pagoda Avenue
Richmond
Surrey TW9 2HQ

Dr K. Browne
Senior Lecturer
School of Psychology
University of Birmingham
Edgbaston
Birmingham B15 2TT

Dr P. Buck
Senior Lecturer in Obstetrics and Gynaecology
St Mary's Hospital
Whitworth Park
Manchester M13 0JH

Professor L. Cardozo
Consultant Obstetrician and Gynaecologist
King's College Hospital
Denmark Hill
London SE5 9RS

Dr L. Cassidy
Consultant Obstetrician and Gynaecologist
Inverclyde Royal Hospital
Larkfield Road
Greenock
Renfrewshire
PA16 9LB

Dr S. Creighton
Consultant Obstetrician and Gynaecologist
University College Hospital
Grafton Way
London WC1

Dr C. Davenport
Specialist Registrar in Public Health
Directorate of Public Health and Health Policy
Oxfordshire Health Authority
Old Road
Headington
Oxford OX3 7LG

Dr M. Doyle
Consultant Obstetrician and Gynaecologist
Arrowe Park Hospital
Arrowe Park Road
Upton
Wirral L49 5LN

Mr C. Kelleher
Research Fellow to Professor Cardozo
The North Hampshire Hospital
Aldermaston Road
Basingstoke
Hampshire
RG24 9NA

Mr F. Lawton
Consultant Obstetrician and Gynaecologist
King's College Hospital
Denmark Hill
London SE5 9RS

Professor D.M. Luesley
Professor of Gynaecological Oncology
Directorate of Obstetrics and Gynaecology
City Hospital NHS Trust
Dudley Road
Birmingham B18 7QH

Dr M. Mann
Senior Registrar in Community Gynaecology
104 Wesley Park Road
Selly Oak
Birmingham B29 5HA

Professor P.M.S. O'Brien
Academic Department of Obstetrics and Gynaecology
North Staffordshire Maternity Hospital
Hilton Road
Harpfields
Stoke-on-Trent
ST4 6SD

Mr A. Parsons
Consultant Obstetrician and Gynaecologist
29 Church Street
Crick
Northants NN6 7TP

Dr E. Payne
Consultant Obstetrician and Gynaecologist
Birmingham Heartlands Hospital
Birmingham

Mr D. Pickrell
Consultant Obstetrician and Gynaecologist
Worcester Royal Infirmary
Newtown Road
Worcester WR1 3AS

Mr C. Redman
Academic Department of Obstetrics and Gynaecology
North Staffordshire Maternity Hospital
Hilton Road
Harpfield
Stoke-on-Trent ST4 6SD

Dr D. Walker
Consultant Obstetrician and Gynaecologist
Royal United Hospital
Combe Park
Bath BA1 3NG

Preface

Common Conditions in Gynaecology: A problem-solving approach is not meant to be a fully comprehensive gynæcological text although the majority of conditions, particularly those occurring commonly, are well covered. The emphasis of this book is in dealing with problems rather than learning pathologies. This book advocates a patient-centred approach to the discussion, diagnosis and management of conditions presenting as gynæcological problems. The emphasis throughout is on the acquiring of clinical skill rather than the storage of facts. The authors acknowledge of course that some factual information is necessary, but only to provide the basis upon which to build the clinical and communicative skills that are now well recognized as the major goals of a medical education. No book can ever replace the experience gained in the clinic and at the bedside, this book aims to complement and indeed foster a true clinical learning mode. We hope that all who have an interest in caring for and learning more about women with gynæcological problems might find this book of value.

Doctor-patient communication and the gynæcological history

1

Kevin Browne and Clare Davenport

1.1 Introduction

Recent discussion on medical practice (Department of Health, 1989a) concludes that clinicians should become more consumer conscious and provide clear information and health advice to patients, the aim being to 'give people a greater individual say in how they live their lives and in the services they need to help them to do so' (Department of Health, 1989b). This reflects a move away from treatment towards prevention, with a greater emphasis on health surveillance and an essential change in the doctor–patient relationship.

Traditionally, clinicians adopt a parental approach to their patients and conventional patients play a 'sick role', showing dependent behaviour and relinquishing any control they have over their health to the doctor. By contrast, the modern approach to clinical practice promotes an equal and interactive process in patient care, whereby the patient is encouraged to discuss their health matters with the doctor and play an active role in decision making. The doctor is expected to respond to the patients' needs by eliciting information, giving advice and instructions, reassuring patients and relatives and passing on information to colleagues. The consultation between a doctor and a patient is a complex process of communication which involves gathering information, making a diagnosis, and explaining to the patient the nature of the condition and its appropriate treatment.

1.2   What is different about gynaecology?

Gynaecology is one of the more difficult specialties to embark upon as an undergraduate in medicine. The close link with the emotional lives of patients is perhaps only equalled in psychiatry. The difference in gynaecology is that verbal exploration of the intimacies of patients' personal and sexual relationships is necessarily accompanied by an intrusive physical examination. Gynaecology is unique in that it only deals with female patients. Thus male doctors are at a distinct disadvantage because they may have more difficulty feeling and conveying empathy and understanding than their female colleagues.

Asking about sexuality is not a part of routine history taking, consequently it may be a source of great discomfort for both doctor and woman. Such questions should not be posed during a physical examination.

Gynaecology also requires doctors to come to terms with personal moral dilemmas and judgemental attitudes when dealing with issues such as termination of pregnancy, infidelity and promiscuity. Tackling these sensitive areas in familiar cultural and social settings is hard enough, but doctors today must also be skilful in dealing with patients of different ethnic backgrounds.

1.3   What is important in the history?

Good history taking is essential to make a diagnosis. As with taking a history in any medical subject a gynaecological history should begin with the concerns of the woman. Besides giving the doctor a basis on which to conduct the remainder of the interview, it reassures the patient by allowing her to communicate what she perceives to be the main problem. Taking a history should not be an interrogation. While initially it will help to have an ordered sequence for the areas to be covered, this style detracts from listening to replies while you think of what question to ask next. Also with experience you will be able to use the replies as prompts for the next question such as 'What sort of family planning are you using?' a reply such as 'Nothing at the moment' might prompt the question 'Are you thinking of getting pregnant?' No prescribed list can accommodate for all contingencies and the art of good history taking is to build enquiry into informal patient-centred conversations. A simple broad outline or logical sequence for history taking is as follows.

(a) Beginning the interview

Greet the patient by introducing yourself. Shake the patient's hand if offered and maintain eye contact. Ask her how she wishes to be addressed. Reduce barriers to communication, such as a desk between the doctor and patient, strong sunlight in the patient's eyes or wearing protective clothing (such as a white coat) inappropriately.

(b) Obtaining relevant information

Ensure privacy and confidentiality. Start with open-ended questions, avoiding jargon. Allow the patient to explain in their own words what their symptoms and their feelings are about the illness. Later, it may be necessary to take control of the interview and focus on vague symptoms by asking direct questions.

(c) Assessing current circumstances

Clarify details of the main problem and its effect on the patient's life and family. How does the patient view the illness? Has the illness changed the behaviour and personality of the patient? How have family members and friends reacted to the patient's ill health? The patient's social environment may have vast influence on how she copes with her illness.

(d) Discussing management plans

Explanations and advice should not normally be given while examining the patient, when the patient may be sensitive only to the procedures of the examination. Before examining, explain to the patient what will be done. When explaining be aware of the patient's expression, the tone of voice, mood and behaviour. These factors (Table 1.1) may give valuable clues to patient understanding and whether the patient feels anxious about the treatment being suggested. Don't feel that you have to fill every silence: silence can be useful when breaking bad news or dealing with sensitive issues, and give the patient an opportunity to express herself.

(e) Ending the interview

The doctor should indicate to the patient that the consultation is coming to an end by summarizing what has been said, avoiding jargon.

Table 1.1 Contributory factors to non-verbal communication

Eye gaze	Eye contact, staring, gaze avoidance, direction of gaze, blinking
Facial expressions	Smiling and frowning, eyebrow movements, colour of the face and cheeks, mouth movements, eye movements
Gestures	Hand and arm movements, hand shakes and nods, other body movements
Non-verbal vocalizations	Tone and pitch, volume, speed, clarity, amount, silence, pauses, hesitations, sighs, laughs
Proximity and touch	Personal distance and bodily proximity, task-related physical contact, touching self (scratching, rubbing, twiddling hair/rings, etc.)
Posture and orientation	Position of limbs/head, position of body, speed and regularity of movement (e.g. walking)
Appearance of patient	Dress, hair, cosmetics, smell, accessories (e.g. cigarettes, bags, spectacles, jewellery)

Allow her enough time to question anything that she doesn't understand or to correct any point she disagrees with in the summary.

1.4   Getting information

A patient-centred interview (Figure 1.1) tends to elicit a great deal
more information and an accepting approach encourages the patient to
be independent. A doctor-centred interview with a barrage of routine
questions, may inhibit communication and the disclosure of personal
information from the patient. A judgemental attitude from the doctor
can often produce feelings of helplessness, dependency and an inability
to cope in the patient.

Attentive listening requires clinicians to suspend their own feelings,
beliefs, values and judgements and give the patient time to think and
express themselves. The doctor should be prepared, appear interested
and be encouraging. Glances at watches, flicking through books or
notes and being interrupted by telephones and people wandering in, is
most off-putting for the patient.

Use 'active listening responses' such as sitting forward, asking questions, giving non-verbal encouragement (e.g. nodding) and paraphras-
ing or summarizing what the patient has said. These cues indicate that

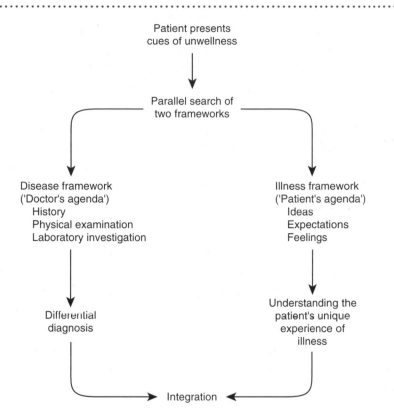

Figure 1.1 The patient-centred clinical interview. From Levenstein *et al.* (1986) *Fam Pract.*, **3**, 24–30. Reprinted by permission of Sage Publications, Inc.

you understand or agree with the information given and emphasize that you are listening carefully. Concentrate on what the woman is saying and try not to interrupt excessively. A few open-ended questions and sensitive responses can save time in the long term and lead to better quality of care.

1.4.1 Unsaid information

During the consultation the effective communicator needs to understand and use verbal and non-verbal communication. By being sensitive to the non-verbal cues of the patient the doctor can become aware of unsaid information (Table 1.1). For example, an awareness of how the woman presents herself.

1. Physical appearance: what does it say about her self-image, health, physical well-being, state of mind? (but beware of making judgements based on your own prejudices!)
2. Manner and bearing: does she appear nervous, defensive, aggressive?

3. Movement and gestures: is she still or fidgety; what is she doing with her hands, feet and head?
4. Posture and stance: is she upright, hunched over, anxious or relaxed?

However, non-verbal signs rarely occur in isolation. Usually they are accompanied by verbal messages which either confirm or contradict the signals. This is a two-way process as the doctor's non-verbal signals can create anxiety or relax the patient.

1.5   The content of a gynæcological history

(a)   History of the presenting complaint

This should include information on duration, change over time, onset, associated events (e.g. a recent birth). Occasionally there will be more than one complaint and there will be a tendency on the woman's part to see them both as part of the same problem. This might or might not be true and it will be important to get the woman to try and prioritize her problems.

(b)   Past medical history

Previous operations and serious illness. Again this might well be directed by the nature of the presenting problem, e.g. infertility might be associated with previous pelvic surgery or pelvic infection and as such these problems might have to be enquired after more carefully. The question should not be 'Have you had any serious illnesses or operations', but 'Have you had any operations on the womb or tubes and do you think you might have had any infection in the tubes; this might have been an episode of pain, fever and a vaginal discharge.'

(c)   Past obstetric history

This should include number of children, ages, birth weights, type of delivery and postnatal problems.

(d)   Terminations and miscarriages

This should include details of when they occurred and, if therapeutic, what methods were used, and were there any physical or psychological problems afterwards.

(e) Drugs

Details should be included of all drugs but especially hormone replacement and oral contraception.

(f) Family history

This should include details of problems, both gynaecological and non-gynaecological, which may be present or have been present in other family members, e.g. family history of cancer or fetal anomalies.

(g) Social history

This should include details of circumstances, the presence of social support (familial and non-familial), occupation of the patient and, if applicable, her partner, and recreational activities including tobacco and alcohol consumption.

(h) General health review

This should include the central nervous system, the cardiovascular system, the respiratory system, the gastrointestinal system and the genitourinary system.

In addition, a gynaecological history has to cover several further topics. These topics can follow the history of the presenting complaint.

(i) The menstrual history

1. Age of menarche
2. Age of the menopause
3. The normal cycle length. This is recorded from the first day of bleeding and regularity should be noted. This information is usually written as the number of days of bleeding over the cycle length, e.g. 5/28.
4. Dysmenorrhoea (pain with menstruation)
5. Menorrhagia (heavy menstrual bleeding). Asking about the number of towels and tampons may give a crude estimate of blood loss but social habits vary. Asking if she has to wear additional protection in bed at night or whether she cannot leave the house during her period are probably better estimates of blood loss.
6. Intermenstrual bleeding (bleeding between what she recognizes as a period)
7. Postcoital bleeding (after or during intercourse).

(j)   Contraceptive history

1.   Current method of contraception
2.   Previous contraception: any problems encountered with previous or present methods, particularly failed contraception
3.   Reasons for changing methods

(k)   Sexual history

This is often the most emotionally charged topic and is usually most appropriately left until the end of gynaecological questioning unless cued otherwise by the patient.

This should include any problems such as:

1.   Pain during intercourse (dyspareunia);
2.   If painful, when, e.g. at the start, during or afterwards;
3.   In cases of infertility the frequency and time in the cycle at which intercourse occurs.

It is also tactful at this point in the interview to enquire about the current relationship between sexual partners if this is relevant.

To assess the patient adequately, prediagnostic life events should be considered in conjunction with their general health (e.g. 'Could you tell me about the events leading up to your illness?'), together with their physical health/psychological well-being/emotional feelings/ personal relationships/lifestyle and personal values/family network/ social supports/intellectual abilities and levels of comprehension.

1.6   Giving information

After establishing a diagnosis, doctors usually explain plans for treatment. At this stage the doctor should identify what the patient wishes to know about her problem, so that she may be informed and reassured. The doctor should lead in giving information as patients may find it difficult to ask for information.

Keep in mind that medical ethics dictates that patients should not take tablets without prior consultation about their purpose and possible side-effects. Patients must be told about any investigations performed upon them, the reasons for doing the investigation and what it will entail. This even applies to the simple procedure of a blood test. Where written consent is not required, verbal consent means that the patient must understand the procedure, therefore allowing informed consent.

It is important to provide information to reduce the patient's anxiety but it may be difficult to assess how much the patient wishes to know. The more traditional approach is to 'play safe' by providing little information to the patient. This can create stress and anxiety and results in low compliance from the patient. However, for a minority of patients too much information may increase stress, so accurate assessment is necessary.

The presentation of information can be improved by:

1. Providing the most important information first
2. Stressing the importance of the information
3. Use of simple language
4. Presenting material in separate categories
5. Repeating information
6. Making advice specific, detailed and concrete
7. Providing information leaflets.

These techniques help the patient to understand and assimilate the information given, and thereby aid coping with the stress of illness and promote patient satisfaction and compliance.

1.6.1   Explaining to patients

To give information the doctor must take account of the person the explanation is aimed at. This means assessing the patient's level of understanding and education and knowledge about medical and health matters. The most important characteristics of explaining are:

1. Clarity
2. Enthusiasm
3. Speaking in short sentences
4. Showing great interest in and logical organization to your explanation
5. Direct speech
6. Relevant to the problem

A vocabulary should be used appropriate to the educational level of the patient with emphasis on important points without vagueness. Sometimes examples and illustrations may be useful.

1.6.2   Breaking bad news

Obviously, being told that you have a serious illness is a frightening experience for a patient. It is especially frightening because the patient feels no longer in control of events. For most individuals, finding out more about the nature of a particular illness that they are experiencing and its likely course of treatment is important in allaying anxiety and in helping to re-establish a sense of control over their lives. However, not everyone copes in this way and medical staff need to take cues from patients about how much information to give and when to give it.

Individuals may have problems communicating to doctors which symptoms are most troublesome or causing them most anxiety. Doctors often focus on aspects of an illness most critical for health and may sometimes give less attention to symptoms such as an irritation or skin rash which is causing a patient considerable distress.

It is easier for patients to have a sense of being in control if the doctor involves them in decisions about their care. Patients should be given the opportunity to express fears about their illness and these fears should be taken seriously. Much of the long-term support will depend on the primary care team and it is important to ensure that they are also informed of the problem, what you have said and how you perceived the woman's reaction.

1.6.3   Coping with angry patients

When breaking bad news to women or their relatives doctors should be aware of the potential for angry emotions. Most individuals feel shock and numbness after being told about a very serious or terminal illness. For some individuals this can turn to angry and abusive behaviour, as they attempt to deny what is happening to them. Under these circumstances it is important for the doctor to behave in a way that will calm rather than escalate the situation.

When dealing with an angry patient, the doctor should remain seated and invite the patient to sit down. The doctor should give the impression of being calm, self-controlled and confident without being dismissive or overbearing. It is advisable to keep talking to the patient in a normal tone of voice, and to acknowledge the patient's anger (e.g. 'It sounds as if you feel very angry with what has happened, would you like to talk about it?').

If the patient fails to calm down, then the doctor should move slowly away, while reassuring the patient.

> Learning points
>
> Diagnoses rely on accurate information
> Accurate information requires effective communication
> Patient centred interviews result in improved patient communication
> Patient centred interviews result in improved patient satisfaction
> Gynæcology covers personal and sensitive areas
> Overcoming embarrassment (patient and doctor) is vital
> Privacy and confidentiality are essential

Further reading

Balint, M. (1968) *The Doctor, His Patient and the Illness*, 2nd edn. Pitman, London.

Byrne, P.S. and Long, B.E. (1976) *Doctors Talking to Patients*. HMSO, London.

Department of Health (1989a) *Working for Patients*. HMSO, London.

Department of Health (1989b) *Caring for People*. HMSO, London.

Myerscough, E.R. (1989) *Talking with Patients: a Basic Clinical Skill*. Oxford University Press, Oxford.

2 The gynæcological examination

Laura Cassidy

2.1 Introduction

The gynæcological examination involves the same clinical skills as required for examination elsewhere, although there are certain specialized parts of the examination such as bimanual examination and passing speculae that will not have been encountered in other disciplines. What really sets the gynæcological examination apart, however, is its intimate nature, which can cause embarrassment in both the patient and examiner. Unless this hurdle can be overcome, information gained from the examination may not be reliable. The woman must be put at ease, and the reason and nature of the examination be explained to her. Male clinicians should always have a female 'chaperone' present at the time of examination.

Clinical examinations should be systematic. First, form a list of possible diagnoses based on the history. Although you should note other abnormalities, your examination should aim to refine this list. Knowledge of the anatomy of the female genital tract forms an essential basis from which to proceed, and will help you to understand and also explain to the patient the nature and natural history of the conditions that may be present.

2.2 Essential anatomical knowledge

2.2.1 The external genitalia (Figure 2.1)

(a) Vulva
The vulva is the general term applied to the external female genitalia. It extends from the mons pubis anteriorly to the perineum posteriorly

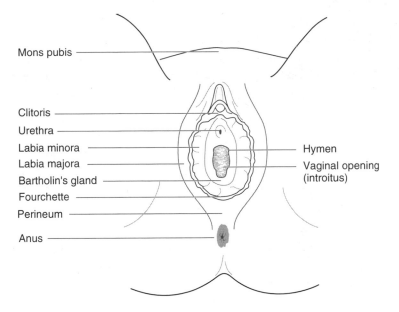

Figure 2.1 External genitalia.

Mons pubis

Clitoris
Urethra
Labia minora
Labia majora
Bartholin's gland
Fourchette
Perineum
Anus

Hymen
Vaginal opening (introitus)

and is bounded laterally by the labia majora. Within these borders lie the labia minora, the introitus (or vaginal entrance), the clitoris and the hymen.

Clitoris   This small slightly bulbous structure lying at the apex of the labia minora is the equivalent of the penis. It contains vascular erectile tissue and is the most sensitive area of the vulva.

Mons pubis   This area overlies the symphysis pubis from which it is separated by a fatty layer. The area is devoid of hair till adolescence. In adult life the surface hair-bearing squamous epithelium contains sweat glands, sebaceous glands and of course hair follicles. It is therefore prone to skin diseases such as sebaceous cysts, boils, warts, psoriasis and squamous carcinoma.

Labia majora   These hair-bearing folds of skin extend from the mons pubis anteriorly to the perineum posteriorly. They are the equivalent of the scrotum. Because they consist of the same structure as the mons they are prone to the same skin disorders. Because at rest the folds are generally in apposition one occasionally encounters equivalent lesions on either side of the labia. In the case of ulcerating squamous carcinoma these are known as 'kissing ulcers'. An infective aetiology is therefore postulated but as yet not proven.

Labia minora   These thin flaps of non-hairbearing skin lie on either side of the introitus, internal to the labia majora. They extend from the clitoris anteriorly where they create a fold or prepuce, to the perineum posteriorly where they join in the fourchette. The urethral opening is situated below the clitoris at the anterior aspect of the introitus between the labia minora.

Hymen   This fine annular membrane surrounds the entrance to the vagina. Initial coitus may cause hymenal tears and slight bleeding particularly if there has been no prior interference such as the insertion of vaginal tampons for control of menstrual discharge. Occasionally the hymenal membrane is intact or imperforate so that menstrual flow cannot be released. These young women then have monthly abdominal discomfort or backache but no menses despite the obvious development of secondary sexual characteristics. The condition is easily recognized by genital inspection. The hymen is seen to be a complete membrane occluding the lower vagina with a bluish discoloration bulging the membrane downwards. The condition is cured by incision into the membrane to release the menstrual discharge and excision of residual hymenal skin tags.

Perineum   This layer of relatively hairless skin and subcutaneous tissue stretches from the fourchette to the anus. It varies in anteroposterior length from 2 cm to 6 cm approximately. It is greatly stretched during the second stage of labour during which it may be torn or cut (episiotomy). The skin lies very close to the underlying muscles of the pelvic floor.

(b)   Pelvic floor (see also Chapters 11 and 12)

This consists of all the structures from the pelvic peritoneum to the perineal skin. Between these lie a fat layer, a muscle layer (levator ani) and a fibromuscular layer (the urogenital diaphragm). Posteriorly lies the anal sphincter. Posterolaterally lie Bartholin's glands.

Levator ani   This muscle group is the main support of the pelvic floor. It stretches from the pubis anteriorly to the sacrum and coccyx posteriorly and laterally to obturator internus and the ischial spines. The muscles join in the midline and are perforated by urethra, vagina and rectum. The levator ani muscle group comprises: (a) pubococcygeus running from pubis to coccyx and with fibres inserted into vagina and rectum; (b) ileococcygeus running from ischial spine to coccyx; (c) ischiococcygeus (or coccygeus) lying behind ileococcygeus and running from ischial spine to lower part of sacrum and upper coccyx. Damage to pelvic floor muscles during childbirth leads to weakness which may cause urinary and faecal incontinence and prolapse of uterus, vagina, bladder and rectum, particularly after the menopause when lack of oestrogens adds to the loss of elasticity of tissues.

Bartholin's glands These are situated posterolaterally one on either side of the introitus. They are less than 1 cm in diameter and produce mucous secretions mainly to help lubricate the vagina and introitus during sexual intercourse. The duct of the gland runs downwards and inwards and opens external to the hymen but internal to the labia minora. The duct may be 2 cm in length and is prone to occlusion particularly with infection. This allows the gland to enlarge causing a cyst or more commonly, if infected, an abscess which may be 10 cm or more in size.

(c) Blood supply, lymphatic drainage and nerve supply

Arteries The vulva is a vascular area supplied mainly by the terminal branch of the internal iliac artery (internal pudendal artery) and particularly anteriorly by the superficial and deep external pudendal branches of the femoral artery.

Lymphatics Lymphatics from both sides of the vulva intercommunicate and drain directly anteriorly to the mons pubis, lower abdominal wall and superficial inguinal nodes. From here they drain to the deep inguinal (femoral) nodes and ultimately to the external iliac nodes. Radical surgery for vulval neoplasia must, therefore, include both sides of the vulva and dissection of the superficial and deep inguinal nodes.

Nerves The skin on the anterior part of the vulva and the mons is supplied by the ilioinguinal and genitofemoral nerves (L1 and L2). The main sensory and motor supply to the pelvic floor comes from the pudendal nerve (S2, 3 and 4).

2.2.2   The internal genitalia (Figures 2.2 and 2.3)

(a)   Vagina

This fibromuscular canal runs upwards and posteriorly between the introitus and the cervix. The bladder lies anteriorly and the rectum posteriorly. The vault (or top) of the vagina has the cervix centrally, anterior and posterior fornices and two lateral fornices. The posterior fornix is larger than the anterior and is the vaginal aspect of the Pouch of Douglas identified from the peritoneal cavity. The vagina varies in length but is longer posteriorly. The vagina is lined with non-keratinized squamous epithelium which is thrown into folds (rugae) allowing it to stretch remarkably at childbirth. Vaginal epithelium is rich in glycogen which is regularly released with cellular turnover. The glycogen is broken down by vaginal commensal organisms liberating lactic acid. This keeps the vaginal pH acidic and acts as an effective barrier to vaginal infection.

Blood supply, lymphatic drainage and nerve supply   The main arterial supply to the lower vagina is from the internal pudendal artery as it is to the vulva. The vaginal artery supplying the remainder may be a branch of the internal iliac or of the uterine artery after it has separated from the internal iliac.

Figure 2.2 Pelvic organs (A-P view).

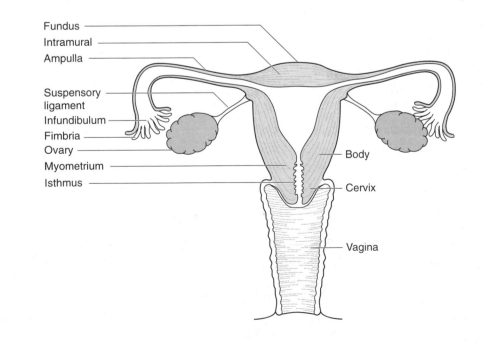

Fundus

Intramural

Ampulla

Suspensory ligament

Infundibulum

Fimbria

Ovary

Myometrium

Isthmus

Body

Cervix

Vagina

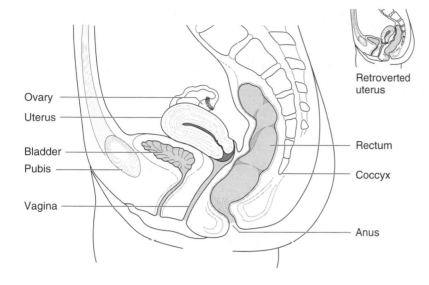

Figure 2.3 Pelvic organs (lateral view).

Retroverted uterus

Ovary

Uterus

Bladder

Pubis

Vagina

Rectum

Coccyx

Anus

The lymph drainage of the lower vagina is to the inguinal nodes as for the vulva. Upper vaginal drainage is equivalent to that of the cervix to internal iliac and para-aortic nodes. Primary vaginal neoplasia is uncommon. The vagina, however, is a not uncommon site of metastases from other organs particularly cervix.

The vagina is remarkably insensitive, its lower part receiving supply from the pudendal nerve.

(b)   Uterus

The uterus is a thick-walled muscular organ shaped like a pear and approximately 10 cm in overall length. The uterus consists of a body (corpus) which projects into the peritoneal cavity, and isthmus which joins the body to the neck (cervix) which protrudes down into the upper vagina.

The body (corpus) has an internal cavity measuring 6–8 cm in length and lined by glandular epithelium (endometrium) which is highly responsive to the hormonal variations of the menstrual cycle. The fallopian tubes are inserted into the uterus through the myometrium at the upper lateral portion of the body known as the cornu. The area of the body above this is termed the fundus.

The isthmus is the small tubular zone joining fundus to cervix. Its main importance is that it forms the lower segment of the uterus in pregnancy.

The neck (cervix) is a tubular organ which forms the lower portion of the uterus. It is 2–4 cm in length and protrudes generally at right angles into the upper vagina. It is made up of connective tissue and involuntary muscle fibres and is lined by stratified squamous epithelium in its vaginal part (ectocervix) and columnar, mucus-secreting epithelium in its canal part (endocervix). Cervical mucus is secreted under the influence of the hormonal changes of the menstrual cycle becoming more profuse at the time of ovulation. The junctional zone between the squamous and glandular epithelium of the cervix is an area of great dynamic change. Metaplasia of cells takes place under the influence of low vaginal pH which itself is dependent on endocrine influences such as female hormones at adolescence, in pregnancy and with combined oral contraceptive preparations. The area between the vaginal junction and any newly created junction is called the transformation zone and is particularly susceptible to infection and neoplastic change. The uterus is normally a mobile organ within the peritoneal cavity. At rest, however, it tends to sit bent forwards in anteversion. The cervix is felt in the mid position in the upper vagina. In about 15% of normal women the uterus is retroverted lying backwards in the pelvis. In this case the cervix tends to be felt high and anteriorly in the vagina (Figure 2.3).

Supports of the uterus   The round ligaments hold the uterine body forwards. They run from the uterine cornu to the internal abdominal ring and thence to the inguinal canal. The uterosacral ligaments run from the supravaginal cervix to the sacrum. The transverse cervical (cardinal) ligaments run laterally from the cervix to the pelvic side walls. The broad ligaments are folds of peritoneum stretching from the side of the uterus to the pelvic side wall. They do not support the uterus but enclose the fallopian tubes in the upper edge, and the ovarian mesentery. The lower part of the broad ligament contains the uterine artery, lymphatic plexus and the distal part of the ureter.

Blood supply, lymphatics and nerve supply   The uterus is supplied entirely by the uterine artery, a branch of the internal iliac artery. The uterine artery runs over the ureter in the base of the broad ligament close to the isthmus of the uterus, an anatomical point of great note to pelvic surgeons wishing to avoid damage to the ureter. The descending cervical branch of the uterine artery leaves the main artery after it has crossed the ureter and supplies the uterine cervix.

Lymphatic drainage from the cervix runs posteriorly in the uterosacral ligaments and laterally in the broad ligaments to the internal and external iliac nodes. The upper part of the uterus drains via the ovarian vessels to the round ligament and inguinal nodes.

The uterus and cervix are quite insensitive. Cautery to the cervix for instance can be undertaken as an outpatient procedure without undue discomfort. However, dilatation of the cervix often causes marked discomfort and occasional syncope. The nerve supply is entirely from the autonomic nervous system running with the main arterial supply.

(c) Fallopian tubes

The fallopian tube is a thin muscular duct which carries the egg (ovum) from the ovary along a ciliated, mucosal lined lumen to the uterus. It is divided into four portions. The intramural part runs through the myometrium into the isthmus, a narrow thick-walled portion leading to the ampulla. The ampulla is wider and thin walled, rather tortuous and the longest portion of the tube. It ends in the infundibulum which is the distended open outer end with fronds (fimbriae) of tissue reaching out towards the ovary.

(d) Ovary

The ovaries are solid ovoid organs about 3–4 cm in length suspended from the uterus on each side by the ovarian suspensory ligaments and attached to the broad ligament by the mesovarium. Primordial follicles are situated in the cortex. Each female child is born with a full complement of follicles.

Blood supply, lymphatic drainage and nerve supply   The ovary and fallopian tube are supplied by the ovarian artery, a direct branch of the aorta. It runs in the broad ligament and reaches the ovary through the mesovarium.

The lymph drainage is directly to the para aortic nodes.

The Fallopian tubes and ovaries are entirely supplied from the autonomic nervous system via the ovarian vessels.

2.3 The examination itself

Most women regard a gynaecological consultation as a most embarrassing experience and therefore approach the visit with considerable fear and apprehension. More than in any other medical situation it is important to provide a quiet private, relaxed environment in which the patient will find herself able to give a full account of her symptoms and history and which will ultimately lead to a relaxed and pain free examination.

Taking the gynæcological history has been dealt with in Chapter 1. Once the history has been taken the doctor should have begun to draw up a differential diagnosis.

2.3.1   Examining the breasts
It is wise to make a thorough examination of the breasts at every gynaecological examination. Active glandular breasts may indicate pregnancy. Also it gives the opportunity to make an early diagnosis of breast neoplasia, particularly in the perimenopausal and postmenopausal woman.

2.3.2   Examining the abdomen
A careful inspection and palpation of the abdomen, including the hernial orifices and groins is essential. The presence of an intra-abdominal or pelvic mass should lead to careful palpation and percussion for hepatomegaly and ascites. (The woman should be asked to empty her bladder to avoid unfortunate errors!)

2.3.3   Positioning the patient
For the abdominal examination the patient should lie flat on her back or, if there is any suggestion of orthopnoea have her head slightly elevated. This is particularly important in the elderly patient. For the pelvic examination the choice of position is often based on the examiner's training. However, you should be aware of the advantages and disadvantages of the positions available.

(a)   The dorsal position
This allows the best opportunity for inspection of the vulva and for bimanual palpation of the pelvic organs. The patient need not move therefore you can easily continue after you have carried out the abdominal examination. It is not the optimum position for detection of prolapse.

(b)   The lithotomy position
The patient lies on her back with the knees or legs elevated in stirrups. This is used for examination under anaesthesia and for visualizing the

cervix at colposcopic examination. This position should be avoided if possible as women feel powerless and immobilized.

(c) Left lateral position (modified Sims)

The patient lies on her left side with the hips and knees flexed. It often helps to have an assistant gently elevate and support the right leg. This position gives good visualisation of the perineum and anus and permits easy identification of a uterovaginal prolapse. The patient may find this position less embarrassing than the dorsal position. It is, however, virtually impossible to carry out a bimanual examination in this position.

2.3.4 Examination of the vulva

The patient should be as comfortable as possible in the position chosen. Adequate lighting is essential. A sheet should be used to cover the abdomen and thighs. A pair of disposable gloves should be worn. Carry out a full inspection of the vulva in the left lateral or dorsal position noting any abnormality including scarring from childbirth. The left lateral or modified Sims position is particularly useful in the case of suspected prolapse. The patient should be asked to bear down or cough to increase pressure on the pelvic floor and help demonstrate a prolapse of uterus or vagina. The insertion of a Sim's speculum (Figure 2.4) and gentle depression of the perineum will allow differentiation between

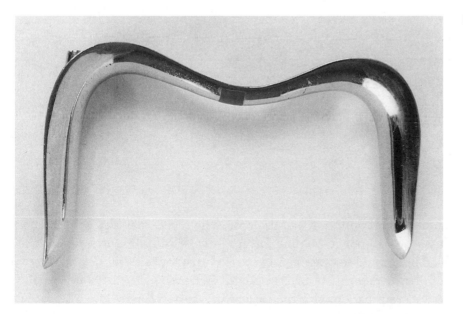

Figure 2.4 Sim's speculum.

Figure 2.5 Sponge-holding forceps.

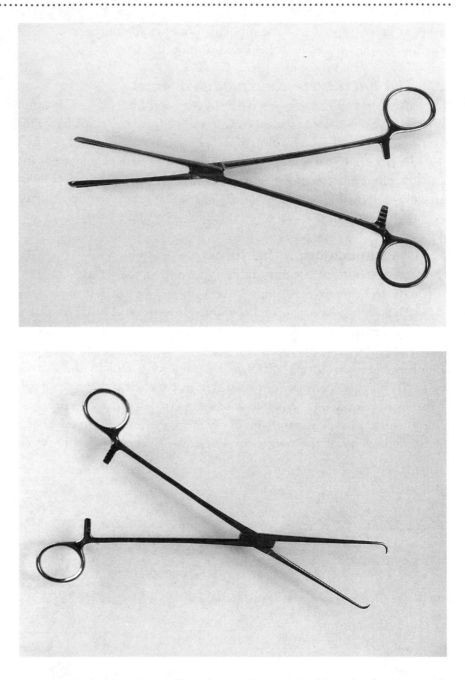

Figure 2.6 Single toothed volsellum.

anterior and posterior wall prolapse. The cervix may also be seen and a tenaculum (sponge-holding forceps or volsellum) (Figures 2.5 and 2.6) can be applied to cause gentle traction thereby determining the degree of uterine descent. In practice many gynaecologists carry out the complete examination with the patient in the dorsal position. The patient has no need to move and the examination may take less time.

2.3.5 Bimanual examination and speculum insertion

The woman is usually asked to take up the dorsal position. Most women are very apprehensive about this part of the examination, so try to put the patient at ease and encourage her to relax. In the dorsal position the patient will have a tendency to plant the soles of her feet firmly on the examining couch and raise her buttocks. This has the effect of tightening the perineal muscles thereby making entry to the vagina more difficult and possibly painful. This can be avoided by asking the patient to abduct the thighs and put the soles of the feet together. It may take a minute extra to achieve this but the time is not wasted. Adequate lubrication on the gloved fingers and speculum is essential but too much lubricant jelly may obscure vaginal discharge and particularly may cause problems with staining of cervical smears. If you insert your finger before the speculum you can assess the size of the vagina especially in the very young or old patient in whom a smaller speculum may be more appropriate. You can also locate the cervix, thereby minimizing trauma when inserting the speculum. Trauma might cause bleeding which could interfere with inspection of the cervix and result in unsatisfactory cervical smears. Most cytology departments prefer the smear to be taken from an untouched cervix. *If a cervical smear is to be taken then the speculum examination is usually performed before bimanual examination.*

(a) Speculum examination

This is performed to inspect the cervix and vagina and is also necessary to take a cervical smear.

Cusco's bivalve speculum (Figure 2.7) is best for inspection of the cervix. It is inserted closed in the antero-posterior position and gently turned through 90° into the lateral position as it is advanced upwards and backwards in the vagina. Gentle pressure on the posterior wall of the introitus prevents disturbing the very sensitive anterior clitoral and urethral area.

Always remember that once you have caused pain to a patient in your gynaecological examination it will be very difficult to gain her confidence sufficiently to allow her to relax to permit further examination. Go gently, paying attention at all times to the patient's reactions to the examination. Once the speculum has been fully inserted the valve should be opened and the cervix visualized. Gentle manipulation will usually allow the cervix to come into view. Observation of discharge, inflammation, polyps etc. should be made and a cervical smear taken if required. The speculum can be gently removed and closed as it is

Figure 2.7 Cusco's speculum.

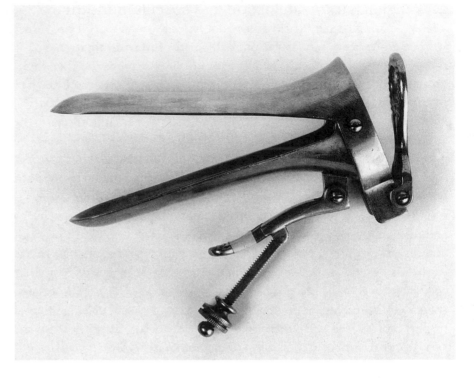

withdrawn past the cervix. Inspect the vagina as the instrument is withdrawn slowly. It is removed downwards and forwards turning gently to the antero-posterior position on withdrawal.

(b)   The bimanual examination

When carrying out the bimanual examination always bear in mind the probable size of the vagina. In young, nulliparous women where the vagina has not been stretched or older postmenopausal women where the vagina may be shrunk and narrowed, perform the initial examination with one lubricated finger. If this is easy a second finger can be gently inserted. In the parous woman a two finger examination should not be difficult. The fingers are inserted up to the cervix. A cervix with neoplasia will often be hard and irregular to the touch, a cervical polyp may be felt. The abdominal hand is of utmost importance in the bimanual examination. Place it below the umbilicus and apply gentle pressure gradually towards the pelvis till the uterus is pushed down on to the vaginal hand. The uterus should be mobile. If it is retroverted the vaginal fingers may be able to manipulate the cervix or alternatively push the uterus out of the Pouch of Douglas through the posterior fornix. Squeezing the uterus is uncomfortable for the patient.

Determine the size, texture, mobility and position of the uterus. It is conventional to describe the uterine size as equivalent to weeks of pregnancy. The vaginal hand should then move to the lateral fornices and the abdominal and vaginal hands gently pushed toward each other. This will allow structures lateral to the uterus to be assessed. Ovarian enlargement can then be determined. The ovaries are mobile organs and may sit high out of the pelvis or commonly, when enlarged, behind the uterus. Distension of the bowel with faeces or gas can be easily confused with an ovarian mass. Ultrasound examination will help differentiate.

When the bimanual examination is complete the vaginal hand should be withdrawn slowly to prevent discomfort. If a pelvic mass is suspected it may be more easily defined by rectal examination. A bimanual examination can be carried out in the dorsal position with the examining finger in the rectum. The Pouch of Douglas can be felt more clearly than on vaginal examination. Another useful technique is to place the middle finger posteriorly in the rectum and the index finger in the vagina. Masses in the Pouch of Douglas and rectovaginal septum such as cancer or endometriosis can be well defined in this way.

Bimanual and speculum examinations	
Inspection	Vulval and/or urethral abnormalities, any prolapse, any incontinence (coughing or straining).
Cusco's speculum	Cervical or vaginal abnormalities, taking a smear, taking endocervical or high vaginal swabs.
Sim's speculum	Cystocele, uterine descent, enterocele, rectocele (seen on withdrawing speculum). (see Chapter 11)
Bimanual examination	Tone of perineal muscles, texture and regularity of cervix, uterine size, position, mobility and tenderness, adnexal (lateral to uterus) masses and/or tenderness, masses behind the uterus.

2.4   Taking a cervical smear

In the United Kingdom cytological examination of the cervix is recommended for all women between the ages of 20 and 65, every three to five years. An active screening programme is in place to encourage all women to be screened regularly in an effort to reduce the mortality from carcinoma of the cervix. Nevertheless many women will still present at a gynaecological clinic having not had a smear for years. The patient should be asked her smear history so that if necessary, a smear can be taken at the beginning of the examination. In general it is better not to take a smear if the patient is menstruating as blood cells will make cytological examination difficult if not impossible.

A clear view of the cervix can be obtained using a Cusco's bivalve speculum with the patient in the dorsal position. The object of smear-taking is to scrape off surface cervical cells and transfer them to a clean dry slide where they are fixed immediately to prevent air drying which distorts the cells. Too much lubricant may distort the staining of the cells. The spatula used is shaped so that its more pointed end is inserted into the cervical os and rotated through 360° scraping gently all round. If any suspicious areas of cervix are seen they should also be smeared. Clear visualization of the cervix is a prerequisite of good smear taking. Scraping the vaginal walls is in general a waste of time.

2.5   Specialized outpatient examinations

In an outpatient setting certain more detailed forms of examination can be undertaken. These include colposcopy, hysteroscopy and endometrial sampling.

2.5.1 Selection of patients

After carefully explaining the procedure, ask the woman about her own feelings about inpatient or outpatient treatment. Outpatient procedures in nervous, uncooperative patients in general fail to provide the information required and rapidly break down the doctor–patient relationship. Elderly patients and the very young are often better examined under a brief general anaesthetic. Endometrial sampling may not be possible in a nulliparous patient.

2.5.2 Colposcopy

Position	Lithotomy
Instruments	Binocular microscope (colposcope) (Figure 2.8a)
	Cusco's speculum (Figure 2.7)
	Endocervical forceps
	Punch biopsy forceps
	Normal saline
	3–5% Acetic acid
	Lugol's iodine
	Monsells solution (ferric subsulphate)
Reasons for referral	Abnormal smear
	Clinically 'suspicious' cervix

(a) Procedure

The cervix is clearly visualized using a binocular microscope (Figure 2.8a,b and Figure 2.9). The squamocolumnar junction (SCJ) is

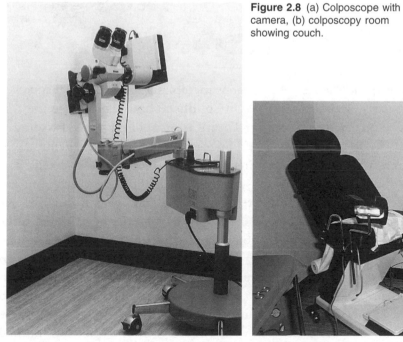

Figure 2.8 (a) Colposcope with camera, (b) colposcopy room showing couch.

(a)   (b)

Figure 2.9 Colposcopic view of the normal cervix (without acetic acid).

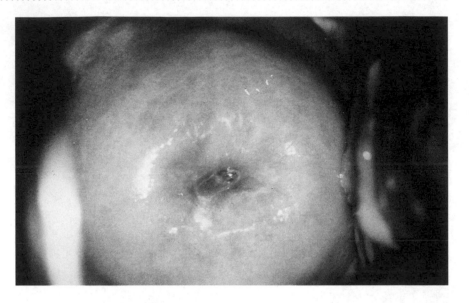

Figure 2.10 Hysteroscope (5mm).

identified and the area of squamous metaplasia to the original squamo-columnar junction inspected. (In childhood and adolescence the SCJ lies well out on the ectocervix.) Dilute acetic acid is applied by spray or cotton wool ball to the cervix. Areas of dysplastic epithelium will become white allowing a directed punch biopsy to be taken. The procedure is only deemed satisfactory if the SCJ can be clearly seen and no acetowhite epithelium extends into the endocervical canal. If the procedure is unsatisfactory a cone biopsy will usually be required. If the histological results of punch biopsy are compatible with the cytology and the colposcopic picture then the treatment can be done on an outpatient basis at a later date by local ablation of the transformation zone using a CO_2 laser, loop or needle diathermy or coagulation with a thermal probe at 120°C. This is best done using local anaesthetic to the cervix after painting the area with Lugol's iodine which stains normal epithelium dark brown. (For further information on colposcopy, see Chapter 9.)

2.5.3 Hysteroscopy

Position	Lithotomy
Instruments	Monocular telescope 3–5 mm diameter (hysteroscope) (Figure 2.10)
	Cusco's speculum
	Single toothed volsellum
	Hibisol antiseptic
	2% Lignocaine
	Uterine sound
	CO_2 insufflator
Reasons for referral	Suspected intrauterine pathology, e.g. intermenstrual bleeding, postmenopausal bleeding, perimenopausal menorrhagia

(a) Procedure

A bimanual examination of the uterus should be carried out before the procedure. Sterile precautions are required. The vulva is swabbed before insertion of a sterile speculum. The cervix is identified, cleaned and grasped with volsellum forceps and a uterine sound (a measuring probe) inserted to measure the length of the uterine cavity. The hysteroscope can then be inserted through the cervix inspecting the mucosa as the instrument is passed. CO_2 insufflation (or saline infusion) distends the uterus and creates a space so that the uterine cavity can be visualized. This allows the uterine cornu and the endometrial lining to be inspected (Figure 2.11). Injection of 10 ml intracervical 2% lignocaine is usually advisable at the start of the procedure as otherwise cervical dilatation can cause pain.

2.5.4   Endometrial sampling

The Pipelle endometrial sampler (Figure 2.12) is a simple hollow tube with an introducer (a bit like a very narrow syringe) which can be passed through the cervix. When the introducer is partially withdrawn

Figure 2.11 Hysteroscopic view of uterine fundus and tubal ostia.

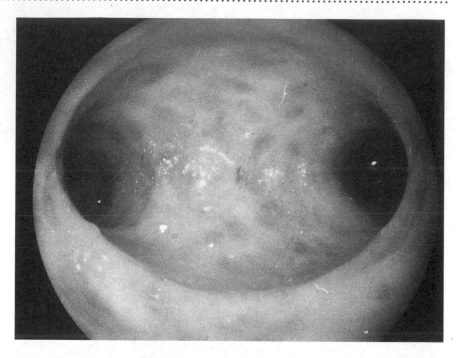

Figure 2.12 Pipelle endometrial sampler.

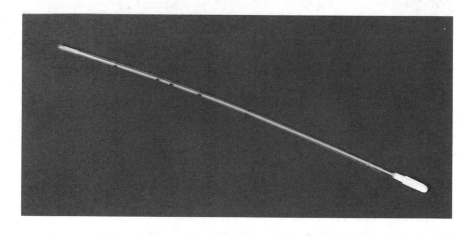

a vacuum is created which sucks the endometrium into the hollow tube. A mechanical pump can similarly be used (Vabra aspirator). Reversing the procedure expels the sample into a pot with fixative. The histological results are excellent. This can be done at hysteroscopy or in a conventional gynaecological outpatient clinic and has significantly reduced the number of inpatient stays for traditional dilatation and curettage.

2.6   Conclusions

The gynæcological examination is an important clinical skill encompassing both basic and specialized techniques. It is vital to consider the

---

Learning points

The examination is directed by the findings from the history

Understand the anatomy

Always ask the woman's permission prior to an examination

With the woman's consent, practise the examination first under anaesthetic

Make sure women are comfortable and given as much privacy as possible

Always look for signs of discomfort

If a smear or swab is to be taken, pass a speculum (Cusco's) before a bimanual examination

Bear in mind the woman's age and parity before inserting fingers or instruments into the vagina

Understand the reasons and selection for more specialized investigations

---

Things you should be able to do

Recognize when you are causing discomfort

Recognize abnormalities of the vulva

Identify prolapse and what is prolapsing

Demonstrate incontinence

Pass a Cusco's speculum and visualize the cervix

Identify common and important cervical lesions such as ectropion, polyps, Nabothian follicles, neoplasia

Take a cervical smear

Palpate the uterus and identify enlargement, retroversion, tenderness and reduced mobility

Recognize masses in the pelvis that are not uterine in origin

Things you should have seen

A cervical smear being taken
A Sims speculum being inserted
An endometrial sample being taken
A colposcopic examination
Hysteroscopy (inpatient or outpatient)
Endocervical and high vaginal swabs being taken

whole patient all the time. Reassurance, explanation and adequate time are essential parts of the process. A solid basis of anatomical knowledge is the starting point from which these skills can be developed. Undergraduates will not necessarily learn the more specialized skills although they should be aware of the techniques and their indications.

Further reading

McMinn, R.M.H. (1994) *McMinn – Last's Anatomy: Regional and applied*, 9th edn. Churchill Livingstone, Edinburgh.

McMinn, R.M.H., Hutchings, R.T., Pegington, J., and Abrahams, P.H. (1993) *A Colour Atlas of Human Anatomy*, 3rd edn. Mosby-Wolfe, London.

Shaw, R.W., Soutter, W.P. and Stanton, S.L. (1996) *Gynaecology* 2nd edn. Churchill Livingstone, Edinburgh.

Snell, R.S. (1995) *Clinical Anatomy for Medical Students*. Little, Brown and Company, London.

Williams, P. (ed.) (1995) *Williams – Gray's Anatomy* 38th edn. Churchill Livingstone, Edinburgh.

Gynæcological emergencies 3

David Pickrell

3.1 Introduction

Most women in their reproductive years are generally fit and healthy. However, gynaecological emergencies do occur suddenly and unexpectedly (Figure 3.1). Excessive and or unscheduled vaginal bleeding or pelvic pain produces anxiety and fear and requires immediate medical care, although not every event is a life-threatening crisis. The accurate assessment of any emergency requires a detailed history, careful physical examination and evaluation of the severity of the underlying condition bearing in mind potential aetiological factors. The woman should be made fully aware of the provisional diagnosis and the therapeutic steps that may be required. Tact and sympathetic handling is essential, an emergency may not only interrupt a woman's normal activities but may threaten and jeopardize hopes of future childbearing and her femininity. This chapter is divided into emergencies associated with complications of early pregnancy and those where a pregnancy is not involved (Figures 3.2 and 3.3), although the two events can occur simultaneously.

3.2 Pregnancy associated emergencies

3.2.1 Diagnosis of pregnancy

The possibility of a pregnancy in any woman in the reproductive age group should always be considered and if necessary investigations such as an ultrasound scan and or a urinary beta HCG (human chorionic gonadotrophin) performed to exclude pregnancy.

Could she be pregnant?

Bleeding or pain following a missed period
Unprotected intercourse
Symptoms of pregnancy: breast tenderness, frequency, nausea
? Positive pregnancy test (self-administered)

Figure 3.1 Gynaecological emergencies: symptoms and signs.

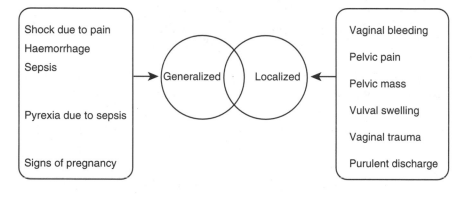

Shock due to pain
Haemorrhage
Sepsis

Pyrexia due to sepsis

Signs of pregnancy

Generalized — Localized

Vaginal bleeding

Pelvic pain

Pelvic mass

Vulval swelling

Vaginal trauma

Purulent discharge

Figure 3.2 The pregnant patient.

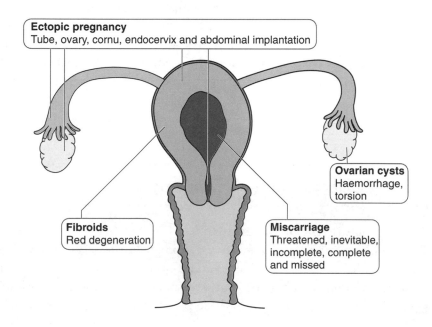

Ectopic pregnancy
Tube, ovary, cornu, endocervix and abdominal implantation

Ovarian cysts
Haemorrhage, torsion

Fibroids
Red degeneration

Miscarriage
Threatened, inevitable, incomplete, complete and missed

Women who state that they cannot be pregnant because they have been sterilized can still be pregnant because the procedure has a failure rate. Furthermore, if a woman conceives following sterilization there is an increased risk that the pregnancy will be ectopic.

Common early pregnancy problems	
Threatened miscarriage	Vaginal bleeding with or without pain. Intrauterine pregnancy. The cervical os is closed.
Inevitable miscarriage	Vaginal bleeding with or without pain. Intrauterine pregnancy. The cervical os is open. The pregnancy is still within the uterus.
Incomplete miscarriage	Vaginal bleeding with or without pain. Intrauterine pregnancy. The cervical os is open. The products of conception are partially expelled.
Complete miscarriage	Vaginal bleeding with or without pain. Intrauterine pregnancy. The cervical os is closed or open and all the products have been expelled.
Missed miscarriage	Vaginal bleeding with or without pain. Intrauterine pregnancy. The cervical os is closed. The embryo/fetus is dead.
Blighted ovum	Vaginal bleeding with or without pain. Intrauterine pregnancy. The cervical os is closed. No fetal parts visible on scan (usually early in first trimester.
Ectopic gestation	No intrauterine pregnancy. The cervical os is closed. Pain and vaginal bleeding may be present.

Although abortion is a well-recognized medical term to define early pregnancy loss it does, however, have quite different connotations for

Figure 3.3 The non-pregnant patient.

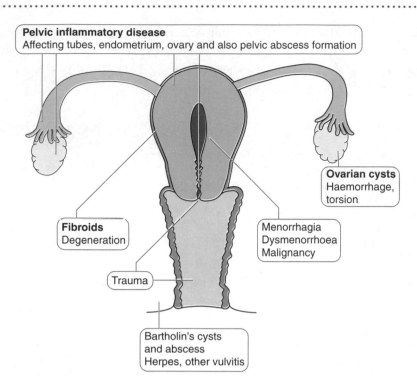

Pelvic inflammatory disease
Affecting tubes, endometrium, ovary and also pelvic abscess formation

Ovarian cysts
Haemorrhage, torsion

Fibroids
Degeneration

Menorrhagia
Dysmenorrhoea
Malignancy

Trauma

Bartholin's cysts
and abscess
Herpes, other vulvitis

the lay public who consider an abortion to be synonymous with a termination of pregnancy. As a result, the term 'miscarriage' will be used continuously in this chapter. Miscarriages occur in approximately 10–15% of all pregnancies although some would suggest that this may be much higher, depending on how such cases are classified.

3.2.2   Diagnosis of miscarriages

(a)   Threatened

A miscarriage is considered 'threatened' when uterine bleeding occurs in the presence of an intrauterine pregnancy before 24 completed weeks of gestation. Bleeding may be variable but pain should be absent or at least minimal. There are usually symptoms and signs of pregnancy before the onset of bleeding, such as a period of amenorrhoea (a missed period), nausea, vomiting and breast tenderness. Abdominal examination should be done first to assess any tenderness or guarding that could indicate an ectopic pregnancy. Bimanual examination will reveal a soft, uniformly enlarged uterus with a closed cervix. The urinary (HCG) assay will be positive and a viable intrauterine pregnancy should be confirmed by ultrasound. It is important to obtain haemoglobin and blood group status. It is equally important to provide a clear and simple explanation of the underlying pathology and possible outcomes. Emphasize that there is no increased risk of congenital abnormalities associated with threatened miscarriages. Bed rest has no proven therapeutic value. The majority of women can be managed quite adequately at home providing that the primary health care team have open access to ultrasonography to determine viability. Hospitalization is not required. However, this group represents a major source for rhesus isoimmunization and blood group status must be assessed in all. If rhesus negative, they should be given 250 international units of anti-D as immunoprophylaxis, preferably within 72h of their first bleed.

(b) Inevitable

This term is used to describe the situation when there is abdominal pain, bleeding and dilatation of the internal cervical os on clinical assessment. The external os may be patulous in multigravid patients, and this should not be erroneously taken as dilatation. An inevitable miscarriage will lead either to a complete or incomplete miscarriage, both of which may be further complicated by sepsis.

(c) Incomplete and complete

A complete miscarriage occurs when the products of conception are passed. Pain settles but slight bleeding may continue for a short time. The conceptus should always be examined to ascertain completeness and to exclude trophoblastic disease. Continual bleeding and pain are suggestive of an incomplete miscarriage and this situation will require hospital care as resuscitation might be required because of heavy bleeding causing hypovolaemia and shock. If the bleeding is very heavy the blood pressure will be low and there will be an associated tachycardia. Urgent resuscitation is necessary with intravenous infusion and type-specific blood. The uterus should be explored and cleared of retained products of conception (uterine evacuation) by curettage under anaesthesia. In pregnancies greater than 14 weeks the process should be expedited by intravenous oxytocin or prostaglandin pessaries to cause uterine contraction. If the woman is shocked but bradycardic a vasovagal effect, due to cervical dilatation, should be considered. Immediate speculum examination of the cervix would reveal the products retained within and distending the cervical canal. Prompt removal with suitable forceps will produce a dramatic response in the patient's condition.

Counselling and explanation should be given to all women following a miscarriage answering the main anxieties of 'why did it happen?' and 'will it happen again?'. Around 80% of spontaneous miscarriages occur because of an abnormal fetus (chromosomal or structural) or abnormal placentation. If three or more miscarriages have occurred then the couple should be reviewed. Aspects of recurrent miscarriage are covered in Chapter 7.

(d) Histological evaluation

The histology obtained from any evacuation of retained products should always be reviewed. If chorionic villi or placental tissue are not seen this may represent the possibility of either a complete miscarriage or an underlying ectopic pregnancy. The latter should seriously be considered if the histology reveals the Arias Stella reaction. If the woman has already been discharged she should be recalled and assessed for the presence of an ectopic pregnancy (see below). Occasionally patients who present at the antenatal clinic will be diagnosed by ultrasonography as having a missed miscarriage. The woman is generally asymptomatic or may still experience pregnancy symptoms. The ultrasound findings may show a gestational sac but no fetus or a sac with a fetal pole but no discernible fetal heart. Explaining missed abortion to mothers is always difficult. In very early pregnancies when this is recognized (blighted ovum) it is sensible to explain that the scan test has failed to detect the heartbeat. This could be because the embryo has stopped growing or because the pregnancy is too small to visualize the heartbeat reliably. Either explanation will still cause distress and a great deal of sympathy and support is required. If there is any doubt, intervention should be delayed for approximately 10–14 days and the scan repeated to exclude the possibility of an error of dating. After a second scan or if a missed abortion can be confirmed after the first scan then the woman should be offered admission to hospital to have an evacuation of the uterus although she may need time to come to terms with this diagnosis.

(e) Septic abortion

Any miscarriage can be associated with sepsis (septic miscarriage). However, criminal interference should be considered particularly if cervical trauma is present. Clinically there may be a profuse and offensive vaginal discharge associated with abdominal and pelvic pain and pyrexia. Infection may produce a local endometritis and para-metritis but may also spread higher up the genital tract to produce a salpingo-oophoritis, peritonitis, septic thrombophlebitis and septica-emia. Renal failure, disseminated intravascular coagulation and septic shock can supervene. Hypovolaemia heralds or accompanies en-dotoxic shock. Rapid resuscitation with large doses of intravenous antibiotics and evacuation of the uterus will reduce the morbidity and mortality. In severe cases, a hysterectomy may need to be considered.

3.2.3 Ectopic pregnancy

If a fertilized ovum implants on any tissue other than endometrium it is termed an ectopic pregnancy. The incidence of this condition is around 1 in 250 pregnancies.

Ectopic gestation: sites

The fallopian tube	Ampulla
	Fimbria
	Isthmus
	Interstitial
The 'uterus'	Rudimentary horn
	Cervix
	Cornu
The ovary	
Other adnexal structures	
Abdomen and pelvis	

Over 90% of extrauterine pregnancies occur within the fallopian tube, 80% of these being in the distal end or ampulla. Non-tubal implantation may occur in the ovary, a rudimentary horn of the uterus, the cervix, the broad ligament or abdominal cavity. Although these sites of implantation are uncommon, they are potentially more danger-ous, with a disproportionately higher mortality. The likelihood of an intrauterine pregnancy and ectopic pregnancy occurring simultaneously in a natural cycle is rare but the risks are increased in women who have undergone assisted fertilization such as gamete intrafallopian tube transfer (GIFT) or *in vitro* fertilization (IVF).

(a) Causes of ectopic pregnancy

Tubal pathology is the major cause of ectopic pregnancies. This may be related to endosalpingeal damage secondary to pelvic sepsis, surgery for a previous ectopic pregnancy and conservative surgery for tubal infertility. Contraceptive practices have a varying effect on the incidence of ectopic pregnancy. The combined oral contraceptive reduces the risk by 90% but the risks increase with the use of the progestogen-only contraceptive (mini pill). Equally, the use of an intrauterine contraceptive device (IUCD) may increase the apparent risk of tubal implantation if a pregnancy occurs as it only prevents intrauterine pregnancy.

(b) Diagnosis

The presentation of an ectopic pregnancy varies: it may be acute or subacute. The classical triad of amenorrhoea, abdominal pain and abnormal vaginal bleeding with cardiovascular collapse represents the acute episode and should immediately raise the possibility of an ectopic pregnancy. The abdomen may be rigid and pelvic examination will reveal marked tenderness together with cervical excitation (pain on moving the cervix laterally or anteriorly). However, it is with the subacute group that the major diagnostic dilemmas occur. Pain is often not severe, bleeding may be irregular with no constant pattern. The only way to diagnose the possibility of an ectopic pregnancy is always to consider it. If there has been intra-abdominal bleeding, either from the distal end of the tube or as a result of rupture, dizziness and fainting may occur. Diaphragmatic irritation may produce subscapular or shoulder-tip discomfort. Blood which has pooled in the Pouch of Douglas may cause pain on defaecation, perineal discomfort or pain, particularly if sitting on a hard seat. The lower abdomen is generally tender and bimanual examination may reveal localized tenderness either laterally or in the posterior fornix. A mass may or may not be palpable. Very occasionally, an ectopic may be silent and discovered during an early antenatal visit, either because of tenderness on bimanual examination or not seeing an intrauterine pregnancy when a scan is done.

> **Common misdiagnoses**
>
> Pelvic inflammatory disease (PID, salpingitis etc.)
> Miscarriage
> Rupture of an ovarian cyst
> Torsion of an ovarian cyst
> Appendicitis

The common pitfalls in misdiagnosis are listed above. Women who may have dysfunctional uterine bleeding and concomitant gastrointestinal pathology can also present with similar symptoms. A raised white cell count is often seen in pregnancy and pyrexia may occur in up to 20% of patients, misleading the clinician to assume that the condition is one of pelvic sepsis.

(c) Investigation

HCG assay and pelvic ultrasound have an important role to play in clarifying a potentially difficult problem.

Assay for the β subunit of HCG HCG is a glycoprotein produced from the syncytiotrophoblast. It is composed of up to two units, alpha and beta, and it is the beta unit that is attributable to the biological activity and its specificity in radioimmune assay. HCG reaches a maximum level of 50 000–100 000 IU/l at 10 weeks gestation and its role is important in maintaining the survival of the corpus luteum. Nowadays, detection of HCG is performed by sensitive monoclonal antibody immunological tests in both maternal urine and blood.

A positive blood HCG assay indicates an ongoing pregnancy or recent miscarriage. HCG levels increase at a definitive rate in normal and ectopic pregnancies. A high (>6000 IU/l) or rising HCG level without ultrasound evidence of an intrauterine pregnancy means that an ectopic pregnancy is a real possibility and in these situations, a diagnostic laparoscopy should be considered (and if an ectopic pregnancy is seen then managed as below).

Pelvic ultrasonography The primary role of pelvic ultrasonography is in diagnosing an intrauterine pregnancy. The identification of a tubal gestational sac occurs in less than 25% of patients with

Figure 3.4 Transvaginal scan of an ectopic pregnancy. Fetal pole visible in sac. Note normal uterus with thickened decidum to the right.

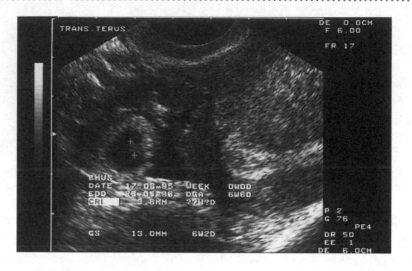

an ectopic pregnancy. Better definition without the need to fill the bladder can now be achieved using a vaginal probe (Figure 3.4). If the woman is in a lot of pain this approach can be difficult.

(d) Management

The management of an ectopic pregnancy depends on assessment of the patient's clinical state. If she is haemodynamically shocked, immediate recourse to resuscitation and surgery is mandatory. In collapsed patients surgery should not be delayed until resuscitation has been achieved.

Where the tube is ruptured and the bleeding profuse, the tube should be removed as quickly as possible (salpingectomy). A diagnostic laparoscopy is not necessary and can add an unacceptable delay in controlling the haemorrhage. Removal of the ovary together with the affected tube should not be considered unless absolutely necessary and never before checking that the other ovary is normal. The chances of an intrauterine pregnancy following salpingectomy for an ectopic pregnancy are reduced to approximately 35% with a recurrence rate in the contralateral tube of 10–15%.

Tubal conservation   A more conservative approach should be adopted in those patients who are haemodynamically stable and where the tube has not ruptured. In these situations the tube can be opened, the pregnancy removed and bleeding controlled. This saves the tube and intrauterine pregnancy rates following are better than after removal of the tube. There is still a future risk of further ectopic pregnancy. The possibility of using a more conservative approach emphasizes the importance of early diagnosis and management. Non-surgical treatments of ectopic pregnancy are currently being evaluated. These include chemotherapeutic agents (methotrexate or actinomycin D) used in the management of trophoblastic disease and are not without major side-effects.

Seventeen deaths, attributable to ectopic pregnancies, were recorded by the Office of Population Censuses and Services (OPCS) between 1989 and 1991 in England and Wales.

3.3 Non-pregnancy related gynaecological emergencies

This section covers emergencies which may result from trauma, infection, haemorrhage, torsion or malignancy and may affect the entire genital tract and occur in all age groups. In the discussion below the major emergency conditions are categorized by which part of the genital tract is mainly affected. This is arbitrary as a single condition, such as infection, may involve the whole genital tract and not necessarily be localized to one area.

3.3.1 The vulva and vagina

Infections of the lower genital tract are not uncommon. The main symptoms are pain, tenderness and dyspareunia. The surrounding tissues will be inflamed and oedematous. These symptoms do not always require emergency admission most being well managed in the community. If, however, acute infections produce severe pain and/or are not responding to primary treatment, hospital admission may be necessary for medical and surgical management.

Emergency problems: vulva and vagina	
Infections	Bartholin's abscess
	Herpes vulvitis
Bleeding	Trauma
Tumours	Bartholin's cysts
	Haematomas
	Malignancies (Rare)
Foreign Bodies	e.g. Retained tampon

(a) Bartholin's cysts and abscesses

Infection of a Bartholin's cyst is a relatively common indication for emergency treatment. The gland and duct are located deep in the posterior third of each labia majora. Obstruction of the main duct causes the gland to become swollen and distended by clear mucoid secretions. Secondary infection causes abscess formation. Primary treatment consists of drainage. The technique is called marsupialization as in effect it opens up a closed cyst forming a sort of pouch. This preserves gland function. If the gland is infected, usually associated with purulent contents, treatment with an appropriate antibiotic is also required. The gonococcus may be an aetiological factor in infection and endocervical, anal and urethral swabs should be taken for culture. If a sexually transmitted pathogen is identified, counselling and contact tracing will be a necessary part of overall management.

In postmenopausal women an enlarged gland should raise the possibility of malignancy and a biopsy is indicated.

(b) Herpes vulvovaginitis

Acute vulvovaginitis, secondary to herpes simplex virus, can produce severe pain and urethral involvement may cause acute urinary retention. The diagnosis is based on the appearance of vesicles, which form flat non-indurated ulcers, 2–5 mm in diameter (Figure 3.5). They are

Figure 3.5 Herpetic vulvovaginitis.

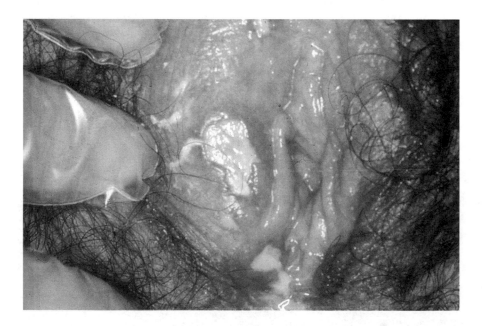

exquisitely tender and painful. Adequate analgesia and acyclovir cream alleviate the symptoms. If urethral involvement prevents micturition, urethral catheterization should be avoided and a suprapubic catheter may be necessary.

(c) Trauma

The female genital tract is protected within the bony pelvis. However, trauma does occur affecting mainly the vulva but sharp objects may penetrate the vagina and surrounding adjacent structures. Vulval haematomas may occur from falling astride a blunt object. Pain may be severe. Small haematomas can generally be controlled relatively easily with pressure and an ice pack. Larger haematomas require surgical incision and drainage and bleeding points may require ligation. Penetrating injuries of the vagina require examination under anaesthesia, superficial injuries are not serious but deep lacerations may involve the bladder, rectum or peritoneal cavity and prompt action is essential. Tetanus toxin should be administered where immunization is in date. However, antitoxin should be considered if evidence of immunization is lacking.

All genital trauma, regardless of severity or cause, is very distressing to women. Such emotional stress, fear, pain and anxiety should prompt an even more sympathetic approach.

The possibility of sexual abuse, particularly in a child or adolescent should be considered in cases of trauma. If there is any suspicion regarding sexual molestation, the most senior clinician should be involved before any examination is undertaken.

3.3.2 The cervix

Haemorrhage from the cervix is primarily related to either cervical malignancy or secondary to cervical surgery. A sudden profuse haemorrhage from the lower genital tract, although uncommon, may be the first presenting symptom of cervical malignancy and should always be considered in older women. Visualization of the cervix would identify a craggy, friable mass which bleeds on contact and is generally associated with an underlying secondary infection. If bleeding is profuse, transfusion may be necessary, particularly in patients who may be anaemic secondary to carcinomatosis. Assessment under anaesthetic should be performed and bleeding controlled by cautery, if possible. The opportunity should be taken to take a biopsy and determine the size and extent of the lesion (staging) with a view to either definitive surgery and/or radiotherapy at a later date. Where infection is suspected, an appropriate broad spectrum antibiotic should be administered.

(a) Iatrogenic haemorrhage from the cervix

With the advent of outpatient management of preinvasive disease, either by diathermy loop excision of the transformation zone or using laser vaporization, secondary haemorrhage occurs in approximately 4% of cases. This may be sufficiently severe to require hospital admission. In the majority of cases, surgical intervention is not necessary. The cervix should be visualized and a high vaginal swab taken and appropriate antibiotics prescribed. It may be possible to either cauterize the cervix with a silver nitrate pencil or alleviate the bleeding by application of Monsel's solution and insertion of a vaginal pack. If the bleeding does not settle then examination should be performed under anaesthesia and any bleeding point sealed either by electrocoagulation or by the insertion of a suture.

(b) Cervical trauma

Cervical trauma *per se* is very unusual and should raise the possibility of criminal intervention in procuring a termination.

3.3.3   The uterus

A sudden unexpected haemorrhage from the uterus is a frightening and potentially dangerous situation. The aetiology may vary from abnormal stimulation of the endometrium, with no underlying pathology, to infection, benign conditions such as leiomyoma (fibroids), endometriosis and malignancy. When haemorrhage occurs without any associated underlying pathology, it is termed dysfunctional uterine bleeding (DUB) (Chapter 4).

Where bleeding is profuse, blood replacement may be required and prompt intervention initiated. An endometrial curettage should be performed to exclude any intrauterine pathology and bleeding can be controlled in the severe case with intravenous conjugated oestrogens. Once bleeding is controlled, progestogen therapy should be added to mature the endometrium and allow a normal withdrawal bleed when treatment is stopped. Management thereafter depends on the age, parity and wishes of the patient.

Emergency problems: uterus

Infections	Endometritis
	Retained products of conception
Bleeding	Dysfunctional
	Fibroids
	Malignancy
Pain	Dysmenorrhoea
	Degeneration or torsion of a fibroid
	Perforated or partially expelled intrauterine device

(a) Management of uterine bleeding

Where haemorrhage is associated with underlying pathology, management is directed at resuscitation if bleeding is acute and severe, followed by assessment of the underlying aetiology and appropriate intervention.

(b) Infectious causes

Infection causing widespread endometritis can be associated with heavy bleeding and is commonly found associated with retained placental tissue, i.e. there will have been a preceding pregnancy. There may be associated abdominal discomfort and pyrexia. Endometrial curettage should be performed and the tissue sent for histology and culture and appropriate antibiotic therapy initiated.

(c) Bleeding due to tumours

Heavy, continuous bleeding secondary to fibroids and malignancy will require surgical intervention with a total abdominal hysterectomy and, in the case of malignancy, combined bilateral salpingo-oophorectomy.

(d) Uterine pain

Low pelvic pain of uterine origin may require admission and may not necessarily be related to abnormal bleeding. Dysmenorrhoea occurs in approximately 50% of women who menstruate and may, in the younger age group, be so severe as to warrant emergency admission. Primary dysmenorrhoea occurs with no discernible organic basis and is related to increased production and release of endometrial prostaglandins during menstruation which results in increased and abnormal uterine activity. Pelvic examination will reveal a normal uterus. This should only be undertaken in sexually active women. In those where intercourse has not occurred, pelvic ultrasonography would exclude any significant pathology.

(e)   Management of dysmenorrhoea

Management is with reassurance and the use of either prostaglandin synthetase inhibitors during the onset of menstruation or by suppressing ovulation with the oral contraceptive, remembering that anovulatory cycles are generally pain-free. Secondary dysmenorrhoea is generally related to pathology within the pelvis such as infection or endometriosis. Occasionally, severe pain may be associated with a submucosal fibroid which is pedunculated, and may result in strong uterine activity, producing extrusion of the fibroid through a dilated cervical os. Occasionally, pedunculated subserous fibroids may undergo torsion and subsequent infarction with acute abdominal pain. Degeneration of a fibroid generally occurs during pregnancy and can mimic an acute abdomen with peritonism, rigidity, nausea and pyrexia with a raised white cell count. Management is generally conservative. Where this occurs without the presence of a pregnancy (Figure 3.6) and where

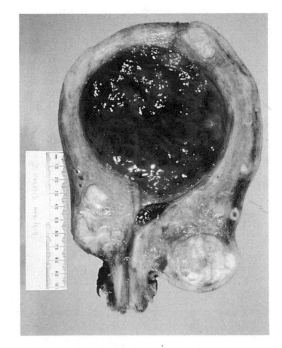

Figure 3.6 Degenerating fibroid.

the woman has achieved desired pregnancies, myomectomy or hysterectomy may be considered.

3.3.4   The adnexae (fallopian tubes, ovaries and parametria)

(a)   Pelvic inflammatory disease (PID)

Aetiology   This occurs as a result of spread of bacteria from the
lower genital tract through the cervix, involving the endometrium,
fallopian tubes, one or both ovaries, the pelvic and possibly abdominal
peritoneum. It is one of the major causes of acute admission to hospital
and appears to be initiated by sexual activity. PID may be subdivided
into two major types. Uncomplicated, i.e. those not associated with
adnexal or inflammatory mass, and complicated, where there is an
association with a tubo-ovarian abscess.

Clinical presentation   Commonly, the presentation is one of abdominal and adnexal tenderness and cervical excitation is frequently
found. There is usually an associated pyrexia and tachycardia with
leucocytosis. The clinical diagnosis of PID is incorrect in up to 35% of
women and a differential diagnosis should include ectopic pregnancy,
appendicitis, ovarian cyst accidents and endometriosis.

Investigation   Laparoscopy should be considered in all women
where the diagnosis is doubtful. Ultrasonography may be helpful in
showing a tubo-ovarian mass. High vaginal swabs, together with
endocervical swabs, should be taken for both routine bacteriology,
chlamydia and gonorrhoea.

  Laparoscopic assessment of the pelvis enables direct visualization of
the uterus, tubes and ovaries and allows access for culture of fluid from
the pelvic cavity. Where the condition is mild, the tubes are
oedematous and reddened but freely mobile with no evidence of
periovarian or peritubal adhesions. In moderate cases, purulent ex-
udates may be seen around the tube and fimbria with the presence of
adhesions which are easily broken down. Where the condition is severe,
there may be an inflammatory complex or abscess.

Management   In the mild cases, appropriate antibiotic therapy, to
include chlamydia, should be started promptly. A cephalosporin and a

tetracycline generally suffice and treatment should be continued for at least ten days. If chlamydia and/or gonococcal infections are identified, the patient's partner should be treated and contact tracing considered. Should the woman be using an IUCD as a form of contraception, this should be removed. Generally, bed rest is advised in the presence of PID but there is no firm evidence to show that ambulation is detrimental.

Pelvic inflammatory disease: sequelae

- Abscess formation
- Pelvic vein thrombosis
- Peritonitis
- Septicaemia
- Infertility
- Ectopic pregnancy

It is important to note that, even in cases of mild pelvic sepsis, infertility is not an uncommon sequelae.

Infertility following pelvic inflammatory disease	
Number of attacks	Percentage infertile
One	11%
Two	23%
Three or more	50%

Not surprisingly, in the presence of tubal pathology, the risks of ectopic pregnancy increase significantly. Where infection is severe and the presence of a tubo-ovarian mass diagnosed, conservative management may be successful. However, should the patient's condition deteriorate or should there be any suspicion that a pelvic abscess has ruptured, a prompt laparotomy should be undertaken. The rupture of an abscess into the peritoneal cavity produces a diffuse peritonitis, septicaemia and septic shock with an associated high mortality. Pelvic thrombophlebitis and ovarian vein thrombosis may occur and are associated with significant morbidity and mortality. Surgery is directed at draining the abscess which can be done either abdominally or transvaginally. Conservative surgery should always be considered in those women who wish to preserve their reproductive function. Where this is not an issue, a complete abdominal hysterectomy with bilateral salpingo-oophorectomy should seriously be considered. During the procedure, the entire abdomen should be examined carefully and adequate drainage of the infected area performed. The likelihood of postoperative ileus in these women is high.

(b) Pelvic mass

It is not uncommon for patients to present with a pelvic mass in association with pain, vaginal bleeding and sepsis. The mass may be easily recognizable on abdominal palpation arising from the pelvis or may be more discrete and found on bimanual examination only. The findings may be due to a fibroid, as discussed above or, indeed, may be suggestive of an ovarian tumour, which is covered later in this chapter.

Should sepsis be present with a concomitant tachycardia and pyrexia, the likelihood of a pelvic abscess should be considered. This may arise, as previously noted, from an ascending infection through the genital tract promoting a tubo-ovarian abscess or, indeed, may be related to a specific bowel pathology, e.g. diverticular disease, Crohn's disease or appendicitis. A detailed history should be taken to determine whether or not there are any pre-existing gastrointestinal symptoms or known gastrointestinal disease. If there is doubt, a surgical opinion should be sought.

Should peritonism be present, resuscitation should be instituted with an intravenous infusion, blood cultures should be taken and a broad-spectrum antibiotic given intravenously. Urgent recourse to laparotomy should be undertaken, in the presence of a general surgeon if bowel pathology is considered. Generally, drainage of the abscess is all that is required and any definitive surgery left until a later date. Intravenous antibiotics should be continued for at least 48 hours, when one would expect to see a significant improvement in the clinical condition. Should there be evidence of persistent sepsis, then it is important to exclude the possibility of a subphrenic abscess and ultrasound, once again, would be the first-line investigative modality.

(c) Problems with ovarian cysts

Sudden onset of lower abdominal pain may be caused by torsion, haemorrhage or rupture affecting an ovarian cyst.

Types of ovarian cysts

Functional ovarian cysts	Follicular
	Corpus luteum
Benign epithelial cysts	Serous cystadenomas
	Mucinous cystadenomas
Benign teratomas	Dermoid cysts
Endometriosis	Endometriomas

In women in their reproductive years, functional non-malignant ovarian cysts (follicular and corpus luteum cysts) are common. These cysts rarely exceed 6 cm in diameter and are generally unilateral and freely mobile (Figure 3.7). Torsion of such cysts is uncommon. However, rupture or haemorrhage into the cyst can occur producing acute lower abdominal pain and peritonism. Such cysts are also associated with menstrual irregularities. Luteal cysts are much larger and bilateral and occur as a result of excess HCG production in association with trophoblastic disease and ovulation induction agents. When a cyst is larger than 6 cm there is greater risk of torsion occurring due to the increased pedicle length. Non-malignant tumours of the ovary are more likely to be associated with torsion than functional cysts. The other common tumour associated with torsion arises from the germ cell and is a benign cystic teratoma (dermoid), the most common tumour of young women. About 15% of these tumours are bilateral.

Figure 3.7 Functional follicular cyst.

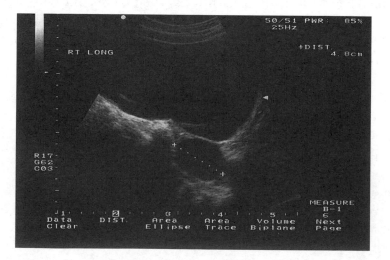

Torsion of ovarian cysts   Torsion of an ovarian cyst produces severe intermittent or continuous lower abdominal pain and is associated with nausea and vomiting. There may be guarding and rebound tenderness and an associated leucocytosis. Abdominal examination may reveal a mass arising from the pelvis and this will be confirmed on bimanual examination. Most ovarian tumours lie behind the uterus or ascend into the abdomen. Dermoid tumours are the exception tending to be anteriorly placed. Benign tumours are clinically unilateral, cystic and mobile with no evidence of ascites whereas, by contrast, malignant growths are solid and fixed (Figure 3.8), possibly bilateral and there may be associated ascites. A differential diagnosis of lower abdominal pain in the presence of an adnexal mass includes ovarian tumours, benign and malignant, endometriotic cysts, inflammatory masses, pedunculated fibroids, ectopic pregnancies and masses associated with the gastrointestinal tract, i.e. diverticular abscess. Diagnostic ultrasonography is helpful in determining the site and size of the cystic lesion and will be able to differentiate between simple and more complex cysts.

Haemorrhage and rupture   In patients where a simple functional cyst has either ruptured or haemorrhaged surgery can be avoided and the patient observed with simple analgesia and reassurance. If, however, there is significant peritonism or a torsion is suspected surgical intervention is necessary. Laparoscopic assessment will provide a definitive diagnosis and haemostasis may be achieved laparoscopically without formal laparotomy. In cysts that have undergone torsion, laparotomy is indicated and management is by removal of the affected

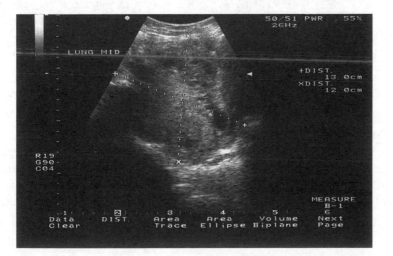

Figure 3.8 Complex solid/cystic ovarian tumour.

adnexa. Bleeding which may not be amenable to treatment laparoscopically requires a formal laparotomy with conservation of ovarian tissue on the affected side. It is imperative to examine the contralateral ovary and, if necessary, biopsy this. Laparoscopic treatment of torsion has been reported but recurrence rates are high.

Emergency problems: adnexae, fallopian tubes, ovaries

Pelvic inflammatory disease   Salpingitis,
    (chlamydia, gonococci, anaerobes)   Pelvic abscess
                                        Peritonitis

Ovarian cysts   Torsion
    (acute presentations)   Bleeding into the cyst
                                        Rupture of a cyst

Occasionally very large cysts can so
    distort the pelvic anatomy as to
    cause urinary retention

Hyperstimulation of the ovaries   With the increasing use of ovulation induction, particularly for assisted fertilization, clinical hyperstimulation is an extremely serious complication, occurring in approximately 3% of cases with 0.8% in the severe form. The risks of severe hyperstimulation are increased if there is a history of polycystic ovarian disease, hyperprolactinaemia and hypothyroidism. Hyperstimulation has occurred in all forms of ovulatory induction, including clomiphene, but is more likely to occur with superovulation using pure FSH or FSH and LH combined.

The World Health Organization recognize three stages of hyperstimulation; mild, moderate and severe. It is the severe form that represents the major worry. The condition is characterized by greatly enlarged ovaries with multiple follicular cysts, stromal oedema and numerous large cystic corpora lutea. There would be associated ascites, pleural effusion, hypovolaemia, oligouria and electrolyte disturbances.

Clinical findings   The patient will present following induction of ovulation. There will be abdominal pain and distension, nausea and vomiting. Hypertension and tachycardia are related to fluid loss from the vascular into the extravascular space, particularly the peritoneal cavity and pleural effusions may result in dyspnoea. Abdominal examination will reveal a tender, distended abdomen with clinical ascites and enlarged ovaries. An ultrasound scan will confirm ovarian enlargement and will enable accurate measurement. Bimanual examination should be performed gently to avoid rupture of the ovarian cysts. In the severe form, the possibility of electrolyte imbalance, coagulation abnormalities, reduced renal perfusion, thromboembolic disease, torsion and possibly intraperitoneal bleeding may occur.

Management   In the mild form, bed rest and analgesia should suffice. However, with severe stimulation, hospitalization and close management is essential. An intravenous infusion should be set up and circulatory volume assessed by means of a central venous pressure line. Intake and output should be carefully monitored, together with daily electrolyte assessment. Diuretics should be avoided since the fluid is in the extravascular space; the use of albumin, however, may improve urinary output. Laparotomy should be avoided unless there is evidence of torsion or intra-abdominal haemorrhage, and ovarian tissue should be conserved if at all possible.

Learning points

Gynæcological emergencies are common and represent a significant part of the average hospital's workload

Many less severe but nevertheless acute gynæcological problems can be and are dealt with adequately in the community

Symptoms and signs may be both widespread (shock) and or local (vaginal bleeding). The whole patient should therefore be assessed.

Always consider the possibility of pregnancy in women presenting with acute gynæcological symptoms.

Resuscitation, identification of the problem, and corrective management with regard to future fertility are the basics of management.

Pelvic inflammatory disease is usually sexually transmitted and is often difficult to diagnose. Prompt effective management and contact tracing will minimize long-term fertility morbidity.

Cases you should try to see

Incomplete miscarriage followed by uterine evacuation
An ectopic pregnancy managed conservatively and by
    salpingectomy
An ultrasound examination for threatened miscarriage
Drainage of a Bartholin's abscess or cyst
Mild pelvic inflammatory disease managed in the community
Severe pelvic inflammatory disease
Counselling following a miscarriage, ectopic pregnancy and
    pelvic inflammatory disease

Further reading

DeCherney, A.H. (1987) Ectopic pregnancy. *Clin. Obstet. Gynaecol.*, **30**, 117.
Houwert-deJong, M.H. *et al.* (1989) Habitual abortion. A review. *Eur. J. Obstet. Gynaecol.*, **30**, 39.

Abnormal vaginal bleeding

4

Shaughn O'Brien and Mark Doyle

4.1 Background

Many thousands of hours are spent in general practice surgeries, gynaecology clinics and operating theatres in the management of problems associated with the ovarian cycle, particularly abnormal bleeding (Figure 4.1).

Only humans and other primates have menstrual cycles. Before contraception was available, and today in many developing countries, frequent pregnancies followed by lactation were the norm and so menstruation was a less frequent occurrence. The life expectancy of women was much less and even if they suffered from problems related to the menopause these were short term and symptomatic; the long-term consequences were not usually of relevance. Selected fertility (the use of contraception and sterilization) significantly increases the number of times women undergo menstruation with its associated problems.

The ovarian hormone cycle is complex, involving subtle and intricate interactions between the ovaries, pituitary, hypothalamus and even cerebral function. Menstruation itself is dependent on cyclical hormone changes affecting the endometrium. However, other systems, particularly the prostaglandins and the various mechanisms which control both local vascular function and coagulation, may significantly alter the menstrual flow and pattern.

Abnormal vaginal bleeding may take many forms: there may be none, it may be infrequent or too frequent. Menstruation may be excessive or perceived to be excessive in amount or duration. A woman may bleed between her periods, following intercourse or she may bleed after the arrival of the menopause. The ovarian hormones and their cycle give rise to several problems not directly related to bleeding, for

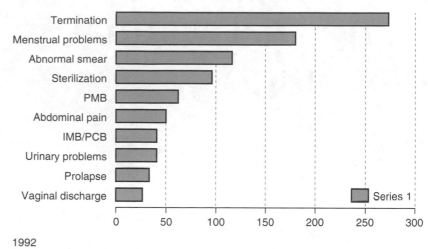

Figure 4.1 Typical referral pattern to gynaecological outpatients in a provincial district general hospital. (Personal communication Mr D.G. Gough.) PMB, postmenopausal bleeding; IMB, intermenstrual bleeding; PCB, postcoital bleeding.

1992
Total number 1019

example painful periods, premenstrual symptoms and cyclical breast pain, fibroids or endometriosis.

Gynaecological and non-gynaecological disease may influence vaginal bleeding. Abnormal bleeding may occur in relation to several complications of pregnancy.

> Points to remember
>
> Whenever a woman bleeds there may be an underlying malignant cause
> Pregnancy is always possible even in the seemingly most unlikely case
> The possibility that this pregnancy may be in the fallopian tube (ectopic)

4.2   Normal menstruation

The arrival of the first period is termed menarche, which means this precisely and no more than this. It is, however, associated with other changes such as the development of secondary sexual characteristics and the emotional changes which herald the arrival of reproductive ability – all of these together are termed puberty.

Menstruation is dependent on the hormonal changes of the ovaries which exert a direct effect on the endometrium. The ovarian cycle is in turn dependent on the pituitary and hypothalamic hormones. An understanding of the endocrinology of menstruation is vital, not only to initiate appropriate investigations but also to explain the problem to women.

4.2.1   Hormonal changes of the normal ovarian cycle

The hypothalamus produces a decapeptide termed gonadotrophin releasing hormone (GnRH). It is produced in a pulsatile pattern throughout the cycle. The GnRH stimulates the pituitary receptors to initiate the production of follicle stimulating hormone (FSH). FSH stimulates the follicles of the ovary to develop.

The follicle has two main functions. First, to produce the ovum ready for fertilization and second, by its granulosa cells, to produce oestrogens. Although several follicles initially develop in the early part of the cycle only one persists, the dominant follicle, the others becoming atretic, shrinking and collapsing.

The oestrogens have many functions through the body but their effect on the endometrium is to give rise to endometrial growth (proliferation). The first 14 days of the cycle are thus known as the proliferative phase. During this phase the endometrium thickens up from basal endometrium with glands and blood vessels elongating and enlarging. When the follicle is the appropriate size, the endometrium at the end of the proliferative phase and the oestrogen levels optimal, feedback to the pituitary changes from negative to positive and a second gonadotrophin is released, luteinizing hormone (LH). This is dramatic and short lived and known as the LH surge. It results in rupture of the ovarian follicle, and hence ovulation, and causes the remaining follicular tissue to become the corpus luteum or yellow body. The yellow appearance is due to infiltration with cholesterol which is a precursor of progesterone. Progesterone is subsequently produced by the corpus luteum under the continued influence of luteinizing hormone.

Progesterone has many actions but its action on the endometrium is to suppress proliferative change, forcing the continually growing vessels and glands to spiral, and to undergo secretory change. Increased glycogen within the glands makes the endometrium suitable for implantation should a fertilized ovum arrive. This is called the secretory phase of the cycle.

If fertilization does occur then the early embryo produces a third, pregnancy specific gonadotrophin, human chorionic gonadotrophin (HCG). This gonadotrophin is structurally similar but functionally identical to LH and thus continues to stimulate the corpus luteum, maintaining and developing the endometrium as appropriate to pregnancy. In this state the endometrium becomes the decidua.

If fertilization has not occurred then the corpus luteum breaks down – luteolysis. The oestrogen and progesterone production of the corpus luteum diminishes rapidly. As the endometrium is dependent on high levels of oestrogen and progesterone, luteolysis causes breakdown of the endometrium and consequently menstruation. The fall in both the ovarian steroids produces negative feedback to the hypothalamus and pituitary once again initiating the release of FSH and thus beginning the next menstrual cycle.

Of importance in menstruation are the prostaglandins which are released in response to the decrease in progesterone. The prostaglandins

Figure 4.2 Endocrine changes in the menstrual cycle.

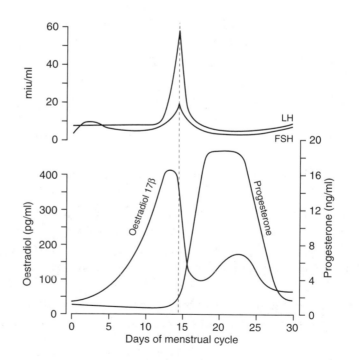

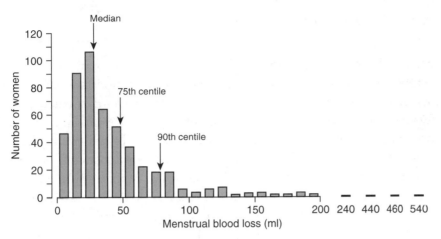

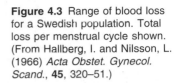

Figure 4.3 Range of blood loss for a Swedish population. Total loss per menstrual cycle shown. (From Hallberg, I. and Nilsson, L. (1966) *Acta Obstet. Gynecol. Scand.*, **45**, 320–51.)

are involved in menstruation itself possibly through vasoconstriction of endometrial arterioles, regulating myometrial contractility and altering local control of haemostasis.

The 'model' menstrual cycle is of 28 days with ovulation occurring on day 14 (Figure 4.2); bleeding lasts for 5 days, with a blood loss of 40 ml in total. The normal cycle can, however, vary in length from 26 to 34 days but the mean is about 28 days. Where cycle length varies the luteal (secretory) phase is normally a constant 14 days, the proliferative phase before ovulation being variable. Bleeding usually lasts from 4 to 6 days though losses from 2 to 8 days can be considered as physiologically normal. The mean volume of loss is 40–50 ml for the whole cycle though it is considered that losses up to 80 ml are within the normal range (Figure 4.3).

The menstrual loss contains more than just variable amounts of blood. Fragments of degenerating endometrium with other serous and cellular constituents are shed. There are chemicals with proteolytic, thromboplastic and fibrinolytic properties that are important as they may influence the amount of bleeding.

The menarche, or the initiation of the first menstruation, is influenced by several factors which include nutrition. Thus the age of menarche has declined in past decades but now in the UK appears to be static at about 12–13 years, most girls beginning between the ages of 11 and 15. Sex hormones are produced, if only in small amounts, before the menarche but the precise hormonal initiator of puberty is unknown.

4.3   Disorders of uterine bleeding

Most of these disorders are problems relating to menstruation. These can be disturbances of amount, frequency, combinations of the two and associated problems such as painful menstruation (dysmenorrhoea). Women will identify most episodes of vaginal bleeding as 'Period' orientated unless it occurs after local trauma, i.e. bleeding after intercourse (postcoital bleeding). Another area where confusion might arise is with intermenstrual bleeding. Women will say that they are having periods every two weeks, yet on careful questioning it will often emerge that one episode of bleeding is consistently shorter and less heavy. Intermenstrual spotting can occur at the time of ovulation and usually only lasts for one or two days. More prolonged intermenstrual loss usually indicates either an endometrial polyp or perhaps early endometrial breakdown. Although not strictly disorders of menstruation, endometriosis and premenstrual syndrome are often considered as associated problems in that the characteristic cyclicity of ovarian function is closely related to these problems.

Abnormalities of uterine bleeding		
Abnormalities of volume	No bleeding	Amenorrhoea
	Excessive bleeding	Menorrhagia
Abnormalities of frequency	Infrequent bleeding	Oligomenorrhoea
	Too frequent	Polymenorrhoea
Painful bleeding		Dysmenorrhoea
Intermenstrual bleeding		
Postcoital bleeding		
Postmenopausal bleeding		

Abnormalities of menstruation are best described clearly in nontechnical language having identified the nature of the problem from the history. For example, it may be more useful to describe a patient as having 'heavy prolonged bleeding' instead of 'epimenorrhagia'.

Common terms relating to menstruation

Menarche	Onset of spontaneous menstruation
Amenorrhoea	No bleeding. Primary when menarche has never arrived or secondary where the periods have begun and subsequently cease.
Menopause	This is physiological and refers only to the permanent cessation of periods at the end of reproductive life. It cannot be defined before the age of 35 though it normally occurs between 45 and 55 the mean being at 51. It cannot be diagnosed until the periods have stopped for 6 months (some authorities say 12 months).
Climacteric	The period of time around the cessation of reproductive life including genital involution
Premature ovarian failure	Permanent cessation of menstruation due to secondary ovarian failure
Menorrhagia	Excessively heavy bleeding at the correct time in the menstrual cycle which may or may not be associated with prolonged bleeding.
Intermenstrual bleeding	Bleeding from the genital tract in addition to the normal menstrual loss; it may be of any quantity.
Breakthrough bleeding	This is technically intermenstrual bleeding but it has less significance if and when it can be clearly related to a precise cause such as the IUCD or the pill.
Postcoital bleeding	Bleeding from the genital tract during or following intercourse.
Postmenopausal bleeding	Bleeding from the genital tract of any amount six months after the last period.
Dysmenorrhoea	Pain associated with menstruation.

4.4   Disorders of volume

4.4.1   Amenorrhoea

No bleeding might also be regarded as a disorder of frequency. Because certain clearly identifiable endocrine and anatomical problems may result in amenorrhoea however justifies its inclusion in this section.

(a)   Primary amenorrhoea

This is considered to be a problem if the menarche has not occurred before the age of 16 years. Primary amenorrhoea is a complex subject. It should be investigated at the age of 16 or sooner if there is also a failure of normal sexual development. If development is normal and the girl does not have an imperforate hymen then further investigation or management is not necessary. It is important to respect the girl and her parents' wishes and at the very least offer reassurance that the majority of normally developed girls will begin spontaneous menstruation before the age of 18 years. If primary amenorrhoea is associated with pain, particularly cyclical, then it is possible that there is an imperforate hymen with cryptomenorrhoea (hidden menstruation) with haemato-colpos and haematometra. That is, blood failing to be released and being retained in the upper vagina and uterus with the risk of retrograde menstruation.

Failure to establish menarche   There are a number of reasons for failing to establish menarche:

1.   The ovary is not stimulated
2.   The ovary cannot respond
3.   The hormones produced are inappropriate
4.   There is no uterus to respond to the appropriately secreted hormones
5.   There is an obstruction

Failure of ovarian stimulation may be due to a pituitary defect such as hyperprolactinaemia or GnRH deficiency. It may, however, be related to weight loss, anorexia, constitutional delay or psychological stress.

The ovary will fail to respond to the pituitary gonadotrophins when there is primary ovarian failure. For example, in Turner's syndrome (45

XO karyotype). When there is mosaicism (46 XO/XX) this may cause confusion because erratic menstruation may have occurred.

Abnormal hormonal secretion will be seen in polycystic ovary syndrome. Congenital adrenal hyperplasia should also be considered. In testicular feminization (46 XY) testes replace the ovaries; testosterone is produced at puberty but the target tissues fail to respond due to androgen insensitivity. These patients are phenotypically female whilst being chromosomally male.

Despite normal production of ovarian steroids, periods cannot be established if there is congenital absence of the uterus due to disorders of Mullerian fusion.

If there is structural blockage of the genital tract below the level of the endometrium then menstrual blood will fail to be released as is seen classically with an imperforate hymen.

(b) Secondary amenorrhoea

Secondary amenorrhoea is the term used for women who have previously menstruated but have stopped menstruating for 12 months in the absence of pregnancy, physiological lactation, hysterectomy or endometrial resection (surgical removal of the endometrium through the cervix whilst conserving the uterus).

Pregnancy must always be considered no matter how unlikely this would seem from the history. The list of other causes of amenorrhoea/oligomenorrhoea is long. The logical way to think of these is by considering the organ at fault and in some ways this will overlap with the causes of delayed menarche.

(c) Causes of secondary amenorrhoea: site of dysfunction

Hypothalamus   Psychological disturbance, weight loss and that due to excessive exercise all produce amenorrhoea through hypothalamic dysfunction. Ballet dancers, air stewardesses and marathon runners commonly experience amenorrhoea.

Pituitary (anterior)   The most common cause is hyperprolactinaemia. There may also be associated galactorrhoea (inappropriate lactation) and, if a large pituitary tumour impinges upon the optic chiasma, a visual defect.

Ovary   The ovary may fail after having previously functioned normally. The most common reason for failure is the menopause which, of course, is physiological. This may occur prematurely.

Another common reason for ovarian failure is secondary hypothalamic–pituitary failure. The two are easily distinguished by measuring the pituitary gonadotrophins which become progressively raised as the pituitary attempts to stimulate the unresponsive ovary.

Hysterectomy, bilateral oophorectomy, endometrial resection, radiotherapy and chemotherapy may also suppress ovarian function. Patients will have noticed these procedures being performed but may not be aware of the association.

Polycystic ovarian syndrome (PCO)   Polycystic ovary syndrome may cause amenorrhoea or oligomenorrhoea. The diagnosis can be confirmed by transvaginal ultrasound (Figure 4.4) as many patients will not have associated hirsutism, obesity and abnormal gonadotrophin levels in their blood (all features of the classical Stein–Leventhal syndrome). There are less common disorders like autoimmune disease and resistant ovary syndrome but these occur rarely.

Uterus   Women who develop secondary amenorrhoea will usually not have congenital anomalies of the uterus and vagina to account for

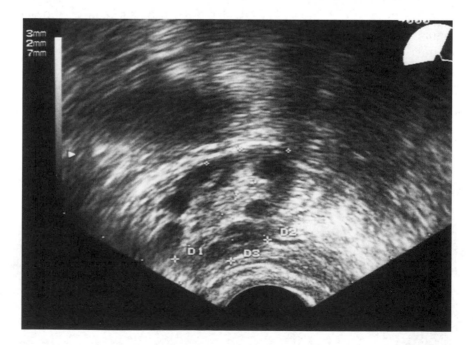

Figure 4.4 Typical ultrasound appearance of the polycystic ovary.

their absent periods. It is, however, possible that they had wrongly interpreted the occurrence of a period and they really have primary amenorrhoea.

Infection (tuberculosis), trauma (Asherman's syndrome), cervical stenosis (cone biopsy) may occasionally be the underlying mechanism of the amenorrhoea.

Common causes of amenorrhoea

Pregnancy
Menopause (physiological and premature)
Constitutional delayed puberty
Weight loss
Exercise
Psychological stress
Hyperprolactinaemia
Polycystic ovarian syndrome (PCO)
Unexplained hypothalamic dysfunction

Although it has been traditional to consider primary and secondary amenorrhoea separately, in practice it is less common to make this distinction rigidly and to consider the two together since often they cannot be separated with any certainty. The amenorrhoea associated with delayed puberty is then dealt with as a separate entity; more precisely it is failure to initiate menstruation or achieve menarche. However, all are complex problems.

Pregnancy is always possible even with primary amenorrhoea.

If a girl has menstruated, but only when taking the oral contraceptive pill or undergoing hormonal therapy, then this is to be considered as primary amenorrhoea.

(d)   Long-term problems with amenorrhoea

Some of these disorders will be associated with important sequelae but in general few will have serious consequences and as such reassurance following thorough investigation is the most appropriate form of management. In other situations, intervention will be required to prevent long-term sequelae.

(e)   Treatment of amenorrhoea

Hyperprolactinaemia   Amenorrhoea due to high serum prolactin levels may be the result of a pituitary microadenoma; this will need full investigation and possibly treatment with bromocriptine or surgery.

Polycystic ovaries   The prolonged unopposed oestrogen exposure of the PCO patient will pose a risk of endometrial cancer. If these patients wish to become pregnant they should receive clomiphene, if not they should receive an oral contraceptive pill containing a non-androgenic progestagen to protect the endometrium.

Premature menopause   Long-term hypo-oestrogenism as is seen with premature ovarian failure increases the risk of developing osteoporosis in later life and also the other effects of the menopause (i.e. increased cardiovascular morbidity). Other women who run similar risks are those who have absent ovaries (testicular feminization) or abnormal ovaries (Turner's syndrome). All these groups will require oestrogen replacement and, if there is an intact uterus, this must be accompanied by cyclical progestagens to prevent endometrial hyperplasia and neoplasia.

In women with testicular feminization the gonads must be removed as these are undescended testes and the risk of malignant change is high.

Imperforate hymen   This can easily be corrected surgically. Other congenital anomalies of the genital tract such as absent uterus and/or vagina are less amenable to surgical correction although with plastic reconstruction women who have a rudimentary or absent vagina can achieve normal sexual relationships although they will be infertile.

(f)   Amenorrhoea: conclusions

It is always necessary to determine the cause of amenorrhoea. Patients want a diagnosis and reassurance or treatment as appropriate. Other considerations include potential fertility, need for contraception, secondary sexual development, secondary sexual characteristics, detection of any potential disorders with serious consequences, e.g. tumour and protection from preventable disorders like osteoporosis and endometrial cancer.

4.4.2 Menorrhagia

Menorrhagia is a common presenting complaint although it may come to light through anaemia and anaemia-related symptoms. Occasionally it may be the manifestation of significant life-threatening underlying pathology such as endometrial carcinoma. It may be the consequence of other gynaecological pathology such as fibroids, adenomyosis, endometriosis, endometrial polyp or pelvic inflammatory disease or it may result from non-gynaecological conditions such as thyroid disease (hypothyroidism) or a coagulation disorder. When no such pathology is identified it is termed dysfunctional uterine bleeding. The precise cause of this is unclear but it may be due to anovulation in some cases with disordered prostaglandin or fibrinolytic function.

It is recognized that many women who report heavy periods actually have average blood loss or less. The patient's history is an unreliable way of determining the diagnosis and, ideally, for a precise diagnosis an objective technique should be employed. Biochemical measurements are difficult and recently a relatively accurate pictorial blood loss assessment chart (PBAC) has been devised to give a more accurate means to quantify the blood loss and thus diagnose menorrhagia (Figure 4.5).

(a) *Identifying the causes of menorrhagia*

To diagnose the underlying cause of menorrhagia the history and examination will give some lead; for instance the triad of menorrhagia, dysmenorrhoea (particularly premenstrual) and dyspareunia may suggest endometriosis. An enlarged uterus found at bimanual examination may suggest fibroids or adenomyosis.

Some causes of excessive menstrual bleeding

Dysfunctional (no recognizable pathology found)
Uterine fibroids
Endometriosis and/or adenomyosis
Chronic pelvic inflammatory disease
Anovulation/metropathic bleeding

Figure 4.5 Simplistic pictorial method to quantify blood menstrual loss. (From Higham, J.M., O'Brien, P.M.S. and Shaw, R.W. (1990) *Br. J. Obstet. Gynaecol.*, **97**, 784–9 published by Blackwell Science, London.)

Tampon	1	2	3	4	5	6	7	8

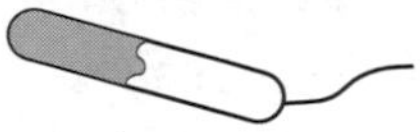								

Clots								
Flooding								

Towel	1	2	3	4	5	6	7	8
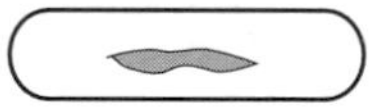								
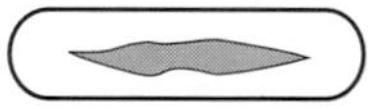								
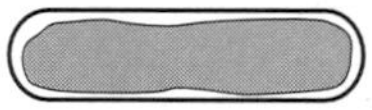								
Clots								
Flooding								

Dilatation and curettage may be necessary to help with the diagnosis and most importantly to exclude endometrial cancer in women over 40. More recently it has become possible to sample the endometrium, as an outpatient procedure, using an endometrial sampling cannula. Endoscopic visualization of the cavity and endometrium may be more appropriate than either of these and can detect pathology otherwise missed. This is hysteroscopy. These are all diagnostic tests and are not treatment methods.

(b) *Management of menorrhagia*

The treatment will depend on the severity of the menorrhagia, the nature of any underlying pathology, the coexistence of other problems, the woman's desire for pregnancy, her age and her personal wishes.

Medical approaches to treatment have centred around hormones, particularly the progestagens; the use of cyclical norethisterone acetate has become very well established with little supportive data. The combined oral contraceptive is highly effective and it is likely that its use will become more widespread, with the newer 'lipid friendly' progestagens. More recently, prostaglandin synthetase inhibitors (mefenamic acid) have been used during menstruation as have various drugs which alter fibrinolytic function (tranexamic acid). Danazol has been clearly shown to be effective in stopping or reducing menstrual loss though there must be an awareness of the potential masculinizing side-effects and the need for barrier contraception.

Recently the progestagen-containing intrauterine contraceptive device has been introduced. This has been shown to reduce menstrual blood loss.

Although it has always been thought that surgical treatment by dilatation and curettage is effective this is almost certainly not true. Hysterectomy is the definitive treatment, but in women who wish to preserve their uterus hysteroscopic endometrial ablation or resection is an alternative. This may be achieved with electrical cautery, laser or microwave generated heat.

Management strategies for menorrhagia

Medical management	Surgical management
Cyclical progestogens	Hysterectomy
Combined oral contraception	Endometrial ablation
Prostaglandin synthetase inhibitors	Laser
Antifibrinolytics	Diathermy
Danazol	Endometrial resection
Iron replacement	(Diathermy)

Figure 4.6 Typical appearance of fibroid.

When there is an identifiable underlying pathology, this will dictate the treatment.

(b)   Menorrhagia due to fibroids

These are common and may exist without causing symptoms (Figure 4.6). They may lie under the peritoneum of the uterus (subserous), within the muscle of the uterus (intramural) or beneath the endometrium (submucous). It has always been suggested that submucous fibroids are most commonly associated with menorrhagia as they increased the surface area of the endometrium and thus the potential area which may bleed. It is not certain that this is true as heavy bleeding is also seen with the other types of fibroids.

Although it is possible to remove just the fibroid (myomectomy) this is rarely the best approach to treatment as the fibroids may recur and the operation is actually of higher risk than hysterectomy. Myomectomy is usually performed by an open operative procedure, but submucous fibroids may be removed via a hysteroscope. Hysterectomy is nearly always the operation of choice though on occasions, particularly when the woman might want children, conservative treatment may be appropriate. It is very rare for fibroids to undergo malignant change.

4.5 Disorders of frequency

4.5.1 Infrequent periods

This is termed oligomenorrhoea. The average cycle length is about 28 days but cycles can be longer and not represent any 'disease' state. Naturally it can and does become a problem should the woman wish to become pregnant as with fewer cycles, her chances of conception will fall.

(a) Perimenopausal bleeding

Periods may also become less frequent towards the time of the menopause as fewer follicles remain in the ovary and these appear to mature less readily as they become resistant to gonadotrophin stimulation.

(b) Polycystic ovaries

Polycystic ovaries are classically associated with oligomenorrhoea and subfertility. It is often the associated fertility problems that cause anxiety in women and this should be considered in any woman complaining of infrequent periods. In the polycystic ovary syndrome there is either absent or very infrequent ovulation.

(c) Anovulation

Occasionally a woman who has been having normal ovulatory cycles will have one or more anovulatory cycles. This situation can result in prolonged oestrogenic stimulation of the endometrium. The endometrium becomes much thicker than normal, the period is delayed and eventually quite heavy and protracted menstrual bleeding occurs. From the woman's point of view her first fear is that she has miscarried (a period of amenorrhoea followed by heavy bleeding with large endometrial fragments). In actuality this is termed metropathic bleeding. Histological examination of the endometrium shows a characteristic 'Swiss cheese' pattern resulting from prolonged oestrogenic hypertophy (glandular hyperplasia).

(d) Effects of contraception

Some women will notice a progressive shortening of the period while using combined oral contraception. Occasionally the periods will stop altogether. This always raises the possibility of pregnancy. A simple pregnancy test can usually resolve the issue and, if the woman is quite happy to continue with her chosen method of contraception, then she should continue with the reassurance that absence of periods will not cause any problems.

A more worrying situation is post pill oligomenorrhoea or amenorrhoea. Women may worry that the pill has caused damage and they will be unable to conceive. Some women will have stopped the pill with the express intent of conceiving. Spontaneous return of normal menstruation is the usual outcome but strong reassurance is necessary and occasionally recourse to drugs to induce ovulation.

Depot progestagen injections are the most likely form of contraception to result in infrequent or absent periods.

4.5.2   Frequent menstruation

It is important to distinguish frequent periods (polymenorrhoea) from normal cycles with intermenstrual bleeding. The latter is due either to pathology such as an endometrial polyp, carcinoma or submucous fibroid or more rarely a cervical neoplasm, or to endocrine problems which remove the normal hormonal support from the endometrium prematurely (luteal phase insufficiency).

Some women will have a short follicular phase resulting in a normal period every three weeks. This does not represent a 'disease', just as a normal cycle occurring every six weeks does not. Furthermore, women do not necessarily maintain the same length of the follicular phase throughout their reproductive lives.

Excessive and more frequent genital tract bleeding is more likely to represent pathology than infrequent bleeding. Frequent menstruation, although not necessarily pathological, can also impose significant limitations on lifestyle and result in a diminished quality of life.

4.5.3   Intermenstrual/postcoital bleeding

Bleeding between the periods is not in itself a danger to a woman. It may, however, be an early sign of underlying malignant pathology. Both endometrial and cervical cancer should be considered possible until proved otherwise by clinical examination, cervical smear and endometrial sampling for histology.

Postcoital bleeding has the same sinister connotation, cervical cancer being the more common association.

Endometrial and cervical polyps as well as cervical ectropion (presence of endocervical glandular epithelium on the ectocervix) may also be responsible for the same symptoms.

4.6   Painful menstruation (dysmenorrhoea)

There is some value in knowing when the problem began in relation to the menarche and when the pain occurs in relation to the bleeding in each cycle. For instance pain that starts for the first time later in life is nearly always associated with gynaecological pathology; this is termed secondary dysmenorrhoea. That which starts early in reproductive life, frequently with the onset of ovulatory cycles, is termed primary or idiopathic dysmenorrhoea. Its name implies that the aetiology is unknown. However, there is now a lot of research to suggest that the prostaglandins are responsible for inducing spasm and ischaemia of the myometrial smooth muscle. The pathological causes of secondary dysmenorrhoea are usually among the commonly encountered gynaecological disorders. These include endometriosis, adenomyosis, pelvic inflammatory disease, sometimes fibroids, the presence of an IUCD and occasionally with cervical stenosis such as may occur after surgery to the cervix (e.g. cone biopsy).

If a young girl presents with painful periods which she has had since the development of her ovulatory menstrual cycles it is unlikely that any investigation is warranted. Treatment can simply be effected by use of the oral contraceptive pill or mefenamic acid. If treatment then fails, investigations such as laparoscopy may become necessary, but this is unusual.

In older women who have developed dysmenorrhoea for the first time then the presence of pathology is more likely, such as endometriosis. Laparoscopy will be required and treatment will be dictated by the nature of the underlying pathological cause.

4.7   Conditions associated with menstruation

4.7.1   Endometriosis and adenomyosis (see also chapter 5)

Endometriosis is the presence of endometrial-like tissue at sites outside the uterine cavity. The common sites are on the ovary, the fallopian tubes, the Pouch of Douglas and uterosacral ligaments. When the ovary is involved there may be large ovarian endometriotic cysts (as they contain old blood and blood products these are sometimes called chocolate cysts). Other areas in the pelvis are frequently involved including the bladder and bowel. In adenomyosis the endometrial type of tissue is found within the myometrium. Medical students often tell me that it may be found in the lung causing monthly haemoptysis whilst forgetting that it occurs in the pelvis!

The most obvious cause of endometriosis would seem to be retrograde menstruation but this cannot be the sole cause as it occurs following tubal occlusion for sterilization. Endometriosis is most commonly associated with fairly specific symptoms such as premenstrual dysmenorrhoea, menorrhagia, deep dyspareunia and infertility.

Management of endometriosis is straightforward. Whether or not treatment is always necessary is controversial. This is particularly the case for small lesions. In these situations there are few data to show that treatment benefits fertility although pain, if present, may be improved.

In simple terms treatment aims to produce a pseudopregnancy or pseudomenopause. In both instances the temporary cessation of menstruation is associated with resolution of the endometriosis in 75–90% of cases.

Danazol and GnRH agonist analogues are the most commonly used drugs. They are used in doses which suppress the cycle completely. Continuous progestagens or oral contraception is occasionally used. There is now increasing use of the new group of drugs, the GnRH agonist analogues which block the pituitary receptors and suppress hypothalamo–pituitary–ovarian function. This, not surprisingly, gives rise to the menopausal side-effects associated with the menopause.

If surgical treatment becomes necessary then this may be simple laser or cautery to the lesions or may entail hysterectomy and oophorectomy.

4.7.2 Premenstrual syndrome

The mechanism of premenstrual syndrome (PMS) is unknown. It is probable that the ovulatory ovarian cycles are the trigger for an unknown psychoneuroendocrine change. The likely contenders for this are differences in serotonin, endorphin or dopamine metabolism. There is as yet no conclusive evidence to support this hypothesis. Equally there seem to be no convincing data to support the theory of a hormone imbalance. The result of these changes is to produce symptoms ranging in severity from the very mild (physiological) to the debilitating. The symptoms may be any combination of physical, psychological or behavioural. They are typically aggression, depression, inability to cope, loss of control, abdominal swelling and cyclical mastalgia.

Premenstrual syndrome is not causally associated with abnormal vaginal bleeding though it frequently coexists. Many different treatment methods have been used but few have been demonstrated superior to placebo.

The oral contraceptive pill, progestagens and progesterone are all commonly used. However, simpler approaches which should be attempted first include reassurance, counselling and relaxation therapy, particularly where the psychological symptoms are most marked. In very severe cases it may be necessary to suppress the endogenous ovarian cycle. This may be achieved with oestrogen implants or patches, danazol, GnRH analogues or (extremely rarely) bilateral oophorectomy and hysterectomy. Unfortunately GnRH analogues can not be used long term because of the menopausal side-effects on bone and the cardiovascular system.

Major surgery can rarely be justified for PMS. However, there are patients who are due to undergo hysterectomy for menorrhagia who also have severe PMS. It would seem reasonable to remove the ovaries in certain of these patients, indeed it may be difficult to justify conservation of the ovaries in severely affected older women.

4.8   Dealing with menstrual problems

4.8.1   Concerns of the patient

Women worry about menstrual problems. If their periods do not start at all then both the girl and her mother fear that there are serious anatomical or endocrine anomalies. If periods then stop there may be worries related to unwanted pregnancy or rare cancers. Bleeding after intercourse, between periods or following the menopause all produce justifiable fears of cancer. This is desirable as certain cancers present early in this way. It is important that women are aware of the significance of menstrual disorders and it is equally important for doctors to know which disorders are important, requiring further investigation and which require simple reassurance.

4.8.2   Relevance of age

Certain problems appear to be related to specific age groups.

1. Around the time of puberty
2. During the reproductive years
3. During and after the menopause

(a) Peripubertal

Delayed onset of menarche and puberty is not uncommon and most often no cause is found. However, there are important causes which require urgent intervention such as an imperforate hymen. There are also many endocrine and chromosomal disorders which require diagnosis or exclusion before reassurance can be given. This type of problem can cause a great deal of anxiety and one needs to be aware of this.

Another particular problem with girls of this age is the importance of the manner in which the initial consultation is made, as this may influence her future relationship with the medical profession. A sympathetic approach is necessary in medicine generally but perhaps here more than any other. Pubertal menorrhagia is common before ovulation becomes established. Once ovulation commences menorrhagia may be replaced by primary dysmenorrhoea, an important and common disorder in this age group. In addition, there are the psychological problems around puberty, worries about sex, contraception, pregnancy and so on.

(b) Reproductive

The established reproductive years are associated with increasing frequency of menorrhagia, endometriosis, pathological dysmenorrhoea, premenstrual syndrome and as the end of ovarian activity approaches the symptoms of the perimenopause and menopause.

(c) Postreproductive

Any bleeding after the menopause requires investigation as malignant diseases of the endometrium and cervix are more prevalent in this group.

4.8.3   Treatment: summary

In most cases the approaches to treatment will fall into the broad categories of surgical and medical and in general the least invasive methods should be the starting point as these will be associated with the lowest morbidity and mortality. A number of questions should be addressed when planning treatment.

1. Is there a risk of giving a certain hormone to a woman in her reproductive years who may be pregnant?
2. Will my treatment be appropriate to a woman who has not completed her family?
3. There are emotional factors which must be considered. Fear of pregnancy or cancer. Emotional changes which may be induced by the hormone therapy itself. Emotional sequelae following loss of her uterus and her biological role of childbearing.
4. Finally and of increasing importance, what is the patient's view of removing her ability to conceive, 'castrating' her or 'depriving her of her womanhood'?
5. Does the degree of abnormal vaginal bleeding or the associated symptomatology justify such intervention?

Learning points

Menstrual abnormalities are common
Pathological causes and exaggerated perceptions of physiological processes can lead to presentation
An understanding of the anatomy and endocrinology is important
Fear of cancer and adverse effects on future fertility are major anxieties among women with menstrual disorders.
Secondary dysmenorrhoea is usually associated with pelvic pathology
Explaining the likely cause is important
Menorrhagia can be managed both medically and surgically
Management is determined by likely cause, age, future fertility.

Further reading

Anon (1986) LHRH analogues in endometriosis. *Lancet*, **ii**, 1016–18.
Boto, T.C.A. and Fowler, C.G. (1990) Surgical alternatives to hysterectomy for intractable menorrhagia. *Br. J. Hosp. Med.*, 44, 93–9.

Higham, J. (1991) The medical management of menorrhagia. *Br. J. Hosp. Med.*, 45, 19–21.

O'Brien, P.M.S. (1987) *Premenstrual Syndrome*. Blackwell Scientific Publications, Oxford.

Shaw, R.S., Soutter, W.P. and Stanton, S.L. (1992) *Gynaecology*. Churchill Livingstone, London.

Shearman, R.P. (1987) Primary amenorrhoea. Secondary amenorrhoea, in *Dewhurst's Textbook of Obstetrics and Gynaecology for Postgraduates* (ed. C.R. Whitfield), Blackwell Scientific Publications, Oxford, pp. 63–79.

Pelvic pain

Pamela Buck

5.1 Introduction

Pelvic pain is a common presenting symptom in gynaecological practice. This does not mean, however, that the cause of the pain is always a disease of the reproductive organs; any medical, surgical or orthopaedic problem causing lower abdominal pain in women may be confused with true gynaecological disorders. Awareness of non-gynaecological conditions is important so that appropriate treatments can be given and inappropriate surgery avoided.

Chronic pelvic pain is an unpleasant condition which causes a great deal of misery to the woman and inevitably has repercussions on partners and family, sometimes even leading to the breakup of a marriage. Conversely, pelvic or lower abdominal pain can be symptoms of a psychogenic disorder initiated or aggravated by marital disharmony. Thus, it is often unclear whether the pain is organic or psychogenic in origin. In either case a sympathetic, patient approach with gentle and tactful questioning is more likely to reveal the true cause of the pain. Sometimes women are reluctant to discuss their anxieties about their relationships with their general practitioner who knows and treats every member of her family. She may find it easier to talk to a gynaecologist whom she perceives as her own doctor and therefore independent.

Pelvic pain, particularly if chronic and persistent, may be due to fear of pregnancy, anxiety about infertility, guilt about sexual indiscretions or fear of cancer.

5.2 Characteristics of pelvic pain

Pain due to gynaecological disease is usually felt in the lower abdomen; uterine pain is hypogastric and ill defined; tubal and ovarian pain is situated above the inguinal ligament. Associated backache is diffuse over the sacrum and may be indicated by the patient rubbing the back of the hand across the sacrum. Pain higher in the abdomen, in the iliac fossa is not usually of gynaecological origin. When pain is related to menstruation it does not necessarily indicate a gynaecological cause, as many other conditions are influenced by the menstrual cycle and the threshold for pain at any site tends to be lower premenstrually. An important feature of pelvic pain is that it may be precipitated or exacerbated by coitus, indeed coital pain (dyspareunia) may be the only symptom and is the most distressing. The relationship of the pain to micturition, defaecation, posture and movement is important in reaching the correct diagnosis. The features of the pain typical of particular conditions are described below in the relevant section.

Pelvic pain: areas for specific enquiry

Relationship with menstruation
Relationship with ovulation (mid-cycle pain)
Sexual intercourse
Defaecation
Micturition
Posture

Gynæcological causes of pelvic pain

Acute pelvic infection
Chronic pelvic infection
Ectopic pregnancy
Endometriosis
Pelvic congestion syndrome
Accidents in ovarian cysts
Pelvic cancer
Degeneration or torsion of fibroids

5.3 Infection

The single most common cause for women developing pelvic pain is infection and although the most common site is the fallopian tubes, other pelvic organs are always affected to a greater or lesser extent. Since the fallopian tubes are in continuity across the uterine cavity and the peritoneal cavity, infection is usually always bilateral. Sexual intercourse is the most common way of acquiring infection with the organisms being directly transmitted in the secretions of the male partner. Other recognized sources of infection are from instrumentation during gynaecological operations, spread from other infected intra-abdominal organs, e.g. appendicitis, or via the bloodstream. Primary infection of the uterus can occur following abortion or parturition.

The following organisms can cause pelvic infection:

- *Neisseria gonorrhoeae*
- *Chlamydia trachomatis*
- Anaerobic and aerobic streptococci
- *Escherichia coli*
- Bacteroides
- Staphylococci
- Pneumococci
- Clostridia species

Swabs taken from the vagina may not yield the relevant organism and it is better to take endocervical swabs and culture for both aerobic and anaerobic organisms (see Chapter 6). Special transport media and culture techniques for *Neisseria gonorrhoeae* and *Chlamydia* should be employed. Positive cultures are likely in acute cases but identification of the organism in chronic infection is unusual.

5.3.1 Clinical features of acute pelvic infection

Infection may produce dramatic illness but may also follow an insidious course with few, if any, symptoms. Many women who present for investigation of infertility and, at laparoscopy, are found to have severely damaged tubes have no record or recollection of such an illness. Acute salpingitis should always be considered when a woman presents with an 'acute abdomen'. She may be very ill, complaining of pain low in the abdomen above one or, more usually, both inguinal ligaments and give a history of fever or rigors. Diarrhoea, a symptom of pelvic peritonitis, may also be reported and may mistakenly lead to a diagnosis of disease in the gastrointestinal tract. Examination usually confirms fever with tachycardia, lower abdominal tenderness with guarding and rebound tenderness. On pelvic examination, the pelvis will feel hot, tenderness is marked and masses representing tubovarian abscesses may be felt. A purulent discharge will be seen issuing from the cervical os, though once an abscess has formed the discharge of pus may cease. Seriously ill women may develop bacteraemic shock with collapse, hypotension and a normal or even subnormal temperature.

5.3.2   Investigation and management of suspected pelvic infection

Laboratory investigations show a normal haemoglobin and a polymorphonuclear leukocytosis. The organism can often be identified from swabs taken from the endocervix.

Severely ill patients should be managed in hospital but in most patients admission will not be necessary and after appropriate microbiological investigations have been done, antibiotics and mild analgesics (if required) can be prescribed. Antibiotic treatment is commenced before microbiological identification is reported. In very ill women, a broad spectrum antibiotic appropriate for *Chlamydia* and gonorrhoea such as erythromycin is given, together with metronidazole to which anaerobic organisms including *Clostridium* are sensitive. The intravenous route is preferred initially and the antibiotic is only changed once the laboratory has determined the sensitivity of the organism. Intravenous fluids are important as the patient may not be able to drink the increased volume required in the presence of fever. Adequate analgesia in the form of pethidine will relieve distress. Surgical intervention is only indicated when the diagnosis is in doubt, for example when acute appendicitis is a real possibility, if the condition deteriorates despite appropriate antibiotics or if more generalized peritonitis supervenes suggesting spreading disease or rupture of an abscess.

5.3.3   Differential diagnosis

Acute salpingitis must be differentiated from other causes of acute lower abdominal pain, appendicitis being the most likely alternative diagnosis. In appendicitis the pain will usually have started in a central position before becoming localized in the right iliac fossa and the findings on pelvic examination are not usually as striking, i.e. unlikely to have 'Cervical excitation'.

Ectopic pregnancy presents with pain but differs in that the pain is unilateral, there is amenorrhoea followed by vaginal bleeding, there is no fever, the white cell count is not usually raised, and human chorionic gonadotrophin is detectable in urine and serum.

5.3.4 Chronic pelvic inflammatory disease

Chronic pelvic inflammatory disease may be preceded by one or more episodes of acute salpingitis, or by repeated subclinical infections. It sometimes presents only with infertility but when pain is present it is situated in the lower abdomen and is often bilateral. It is described as aching or gnawing and is sometimes accompanied by sacral backache. The pain may be worse just before and at the onset of menstruation and dyspareunia is an important feature. Menstrual disturbances result from uterine and ovarian involvement. Other symptoms include tiredness, lack of energy, poor appetite and infertility.

(a) Clinical signs

Physical signs are variable, ranging from none to generalized pelvic tenderness, reduced mobility or fixation of the pelvic organs including retroversion of the uterus. Although hydrosalpinges are a feature of this disease they are rarely palpable and the only abnormality felt may be thickening of the appendages. Inspection of the pelvis at laparoscopy or laparotomy will reveal adhesions between the uterus, tubes and ovaries and adjacent organs such as sigmoid colon, rectum, caecum and small bowel. Adhesions may obscure the pelvic organs completely. When visible the tubes are obstructed and hydrosalpinx formation common (Figure 5.1). Antibiotics and analgesics are useful in the treatment of acute exacerbations, but the effect is temporary.

The treatment of choice is surgical. As the pelvic anatomy is so distorted, total abdominal hysterectomy and bilateral salpingectomy is the only treatment associated with long-term relief of pain. This often poses major problems for the woman and her doctor as many women will want to preserve any chance of fertility. Without new techniques like *in vitro* fertilization, the prospect of these women becoming pregnant is very poor. Very careful counselling and support are necessary prior to any surgical management in these women.

5.3.5 Pelvic tuberculosis

Pelvic tuberculosis is uncommon in the United Kingdom nowadays, most cases occurring in women who have arrived from parts of the world where the disease is endemic, the Indian subcontinent and

Figure 5.1 Tubal hydrosalpinx, a feature of pelvic inflammatory disease.

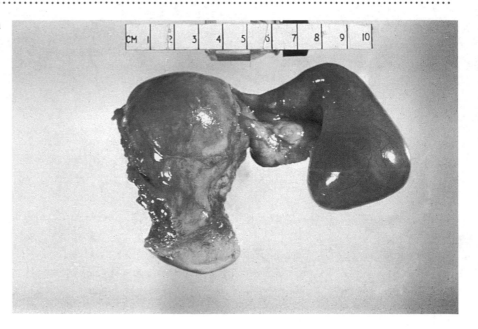

Africa. The symptoms are of chronic pelvic pain, infertility and weight loss, but some present with a more acute illness similar to acute salpingitis. There is evidence of tuberculous infection elsewhere in the body, usually the lungs. Unlike other forms of pelvic infection, the menstrual disturbance is scanty or absent menstruation rather than menorrhagia. Surgery should not be carried out while the disease is active and a full course of antituberculous chemotherapy should be given. As with other forms of the disease, close contacts should be screened by chest radiographs, sputum cultures and Mantoux testing. Spontaneous pregnancy following pelvic tuberculosis is rare and when it does occur is invariably ectopic because if the tubes have remained patent, they will usually not function normally and ovum transport along the tube will not be normal.

5.4   Endometriosis

Endometriosis is a curious disease which has been recognized for over a century. There is a proliferation of endometrium in sites other than the uterine mucosa (endometrial deposits within the myometrium is termed adenomyosis). The cause is still unknown though many theories exist. Implantation of endometrial tissue which has reached the pelvic peritoneum by retrograde menstruation is considered the most likely. Since this last phenomenon is common yet few women develop endometriosis, it has been suggested that affected individuals have a defect of cell-mediated immunity. Another theory postulates that the cyclical bathing of pelvic peritoneum by follicular fluid at ovulation inhibits proliferation of endometrium and hence some women with endometriosis are infertile because their follicles fail to rupture. The disease most commonly occurs in the fourth and fifth decade of life in women who have delayed their childbearing for any reason.

When assessed at laparoscopy or laparotomy the ovary is the commonest site for endometriosis where its appearance ranges from small blue–black spots (Figure 5.2) to cyst formation (Figure 5.3). These cysts are called 'chocolate' cysts because they contain altered blood which resembles chocolate sauce in colour and consistency. Endometriosis also commonly occurs over the peritoneal lining of the pelvis, the Pouch of Douglas and the uterosacral ligaments. Involvement of the serosal surfaces of adjacent organs such as bowel and bladder is not unusual though penetration to the lumen is rare. Endometriosis has been seen in normal women when laparoscopy is performed for other reasons and has been noted to disappear without treatment implying that symptoms and progressive disease are not

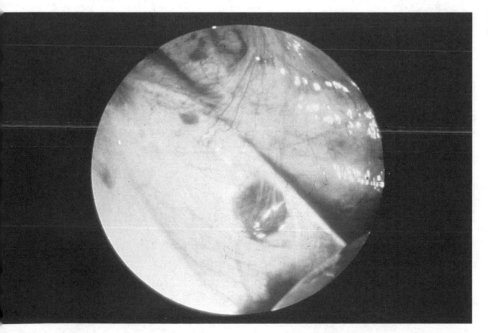

Figure 5.2 Endometriosis – early lesion.

Figure 5.3 Endometriosis – cyst formation.

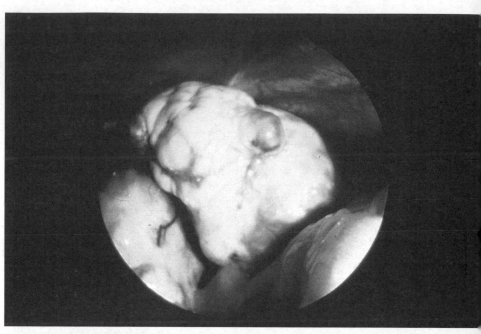

Figure 5.4 Endometriosis: progressive disease. Adnexal adhesion.

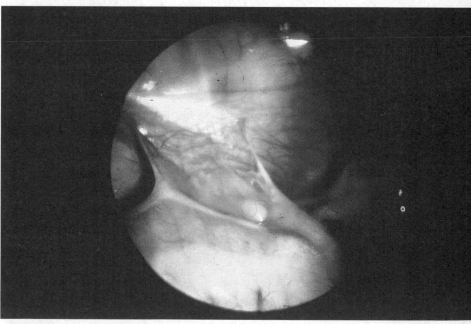

invariable consequences. The characteristic feature of progressive disease is adhesion formation (Figure 5.4). Tubal damage is secondary to adhesions rather than primary endometriotic deposits. The adhesions are dense with scarring of the surrounding peritoneum and in advanced disease the anatomy is grossly distorted and function seriously impaired.

5.4.1 Symptoms of endometriosis

When symptoms do occur the correlation between their severity and the extent of endometriosis is poor, a few spots of endometriosis can produce symptoms just as disabling as cases where the entire pelvic contents are fused together. Dysmenorrhoea typically occurs at the onset of bleeding and lasts for the duration of menstruation, unlike primary dysmenorrhoea which begins a few hours before menstruation starts and lasts for a day. Dyspareunia is explained by the presence of endometriomata in the Pouch of Douglas or in the rectovaginal septum and fixation of the uterus in retroversion. The pain of endometriosis is usually aggravated by menstruation. Pain or bleeding during defaecation or micturition at the time of menstruation suggests bowel or bladder involvement. Rarely, distant sites such as the pleura are involved giving symptoms of cyclical pleuritic pain or even haemoptysis. Implantation of endometrium in the wounds from gynaecological or obstetrical operations gives rise to a mass in the scar which increases in size and tenderness at menstruation.

On abdominal palpation ovarian cysts arising from the pelvis may be felt, though the size rarely exceeds 10 cm and most cysts are confined to the pelvis. Examination of the pelvis may be negative in minimal disease but typical signs are of tender nodules in the uterosacral ligaments palpated through the posterior fornix. As the disease becomes more extensive adnexal masses are palpable, the uterus becomes fixed in retroversion and the posterior fornix appears tethered to the back of the uterus. Simultaneous rectal and vaginal examination may reveal a tender mass in the rectovaginal septum. Tenderness is most marked when examination is carried out during menstruation. If there is doubt about the diagnosis or it is unclear whether the patient is suffering from chronic pelvic inflammatory disease or endometriosis, laparoscopy should be performed when the typical appearances of endometriosis are obvious.

Symptoms of endometriosis	Clinical signs
Dysmenorrhoea	Pelvic tenderness
Dyspareunia	Tender, retroverted uterus
Pain on defaecation	Nodules in uterosacral ligaments
Menstrual disturbances	Adnexal masses
Infertility/subfertility	

5.4.2   Treatment of endometriosis

Before starting treatment it is important to understand what the woman most wishes. This is particularly relevant with regard to fertility. Endometriosis is a difficult concept for untrained personnel to understand and a clear simple explanation should be the first step in management. Any disease that has a 'spreading capacity' raises fears of cancer and this must be tackled early on. Phrases such as 'This condition is not a cancerous disease' should be included in the early explanation. Information leaflets to back up verbal information are helpful as are self-help groups.

Endometrium, wherever situated, is responsive to sex steroids and one of the first clinical observations of this disease was that it improved during pregnancy and after the menopause. Conservative treatments exploit these properties. A pseudopregnancy can be induced by giving a combination of oestrogen and progestagen or a progestagen alone continuously over six months. A combined oral contraceptive containing 30 µg of oestrogen and a potent progestagen is a convenient way of achieving this. A progestagen such as dydrogesterone can be given alone although fluid retention, breast tenderness and nausea might cause discomfort, especially early in the treatment. Androgens also suppress endometriosis but the risks of virilization mean that they are not safe to use. However, a weak androgen, danazol, is effective as it suppresses GnRH production without stimulating the endometrium. Some androgenic effects such as acne, weight gain, muscle cramps, hirsutism and, rarely, voice change, do occur with high doses given over long periods. These are unacceptable to many women and long-term use of danazol can be very difficult. GnRH analogues completely suppress ovarian activity (like the menopause) and cause endometrial atrophy. These drugs are given as a nasal spray or as a subcutaneous injection or depot preparation. As no oestrogen is produced, women notice menopausal symptoms such as hot flushes and vaginal dryness. Calcium loss from bone does occur but is usually not significant if the treatment is limited to six months.

Although medical, conservative treatment can have a dramatic effect on symptoms, they often return once the treatment has been stopped. Medical treatment of endometriosis does not significantly improve the woman's chances of getting pregnant. Surgical treatment is more likely to produce long-term relief of symptoms. Destroying individual lesions using diathermy, or more recently laser, is preferable when the disease is limited and the woman wishes to preserve her fertility. Chocolate cysts may be enucleated while trying to conserve as much normal ovary as possible. If the disease is extensive it can be difficult to conserve the woman's fertility. Total abdominal hysterectomy is probably the best option in women who do not wish to have further children and it is usual to remove the ovaries at the same time as continued oestrogen production might stimulate residual areas of endometriosis. In young women keeping at least one ovary is desirable as long as it is not involved with endometriosis. *In vitro* fertilization offers a chance of pregnancy in women with endometriosis, medical treatment being used to control pain symptoms before and between IVF treatment cycles.

5.5   Pelvic congestion

Pelvic congestion may be mistaken for irritable bowel syndrome, but the important distinguishing feature is the absence of disturbance of bowel habit. The cause of pelvic congestion is unknown. The veins draining the pelvic organs are thin-walled and prone to chronic dilatation with stasis resulting in vascular congestion. This phenomenon varies at different times of the cycle, being worse before a period, suggesting that the sex steroid hormones are involved. Symptoms can be relieved when ovarian function is suppressed. Sufferers are aged between 20 and 40 years with a peak at 30 years. They complain of a dull aching pain in the iliac fossa, dyspareunia and dysmenorrhoea. About half the women who have pelvic congestion also have heavy periods. In common with irritable bowel syndrome scars from previous surgery are commoner than in the general population. Tenderness of the ovaries on pelvic examination and tenderness in the iliac fossa may be the only physical signs. Diagnosis is made on the history and the physical signs but pelvic congestion may be seen at laparoscopy as hyperaemia of the pelvic organs with large dilated veins especially in the broad ligament. Venography, carried out in a large study of this condition, has shown dilatation, pooling and delayed clearance of contrast medium from the pelvic veins. However, venography is too invasive a procedure for routine diagnostic use and has been replaced by ultrasonography.

Medical treatment depends on the suppression of ovarian function using progestagens given orally over a period of several months.

Improvement is gradual and treatment should not be abandoned after one or two months when dramatic results have not been achieved. Oophorectomy should only be carried out as a last resort.

5.6   Other gynæcological conditions

5.6.1   Ovarian cysts

Ovarian cysts do not cause pain unless an accident such as torsion, rupture or haemorrhage occurs. All three cause acute pain. In torsion of the ovary or a cyst the pain may be intermittent over a few hours as the ovary undergoes partial torsion and returns to its usual position before finally undergoing complete rotation, cutting off its blood supply. The history then is of pain of acute onset situated low in the abdomen on the affected side, often associated with vomiting. A further discussion of ovarian cyst accidents is given in Chapter 3.

5.6.2   Ovarian cancer

Cancer of the ovaries causes abdominal distension and disturbed bowel habit. Pain is not commonly associated with ovarian cancer. When pain is part of the ovarian cancer symptom complex, it usually develops more gradually over a few days or weeks. In advanced ovarian cancer, and most women do present with advanced disease, there are usually other features such as malaise, anorexia, weight loss and abdominal distension due to tumour and ascites.

5.6.3   Uterine fibroids

Fibroids are common and are not usually associated with pain. When you explain to a woman that she might have fibroids she will believe that they are the cause of pain. There are, however, some situations where either twisting of a fibroid (if it is on a stalk) or degeneration of the fibroid (hyaline, fatty and red degeneration) cause pain. Red degeneration of a fibroid occurs in pregnancy as a result of venous congestion. Sarcomatous change in a fibroid may present with pain (Figure 5.5).

Anything that can cause stretching of the peritoneum covering a fibroid (i.e. rapid increase in size) will cause pain.

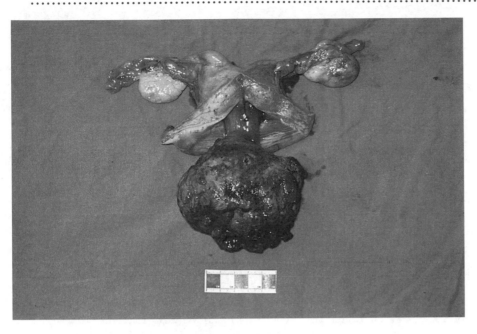

Figure 5.5 Malignant fibroid polyp.

5.6.4   Surgical causes

Surgical causes of pain are seen in gynaecology departments and vice versa. The most common acute surgical condition which is confused with gynaecological disease is appendicitis. The typical features of vague periumbilical pain followed by more localized right iliac fossa pain, nausea, anorexia, fever and leucocytosis do not give rise to confusion, but in pelvic appendicitis the abdominal signs may be absent and tenderness on rectal or vaginal examination is more marked. If there is doubt about the diagnosis a laparotomy or laparoscopy should be performed, in case generalized peritonitis develops or acute salpingitis is missed and treatment delayed.

Constipation is a common condition which may present with abdominal pain, distension and faecal masses may be confused with ovarian cysts or tumours. The abnormal bowel habit will be apparent if the appropriate questions are asked. The rectum may be full of hard faeces. A straight radiograph of the abdomen taken as part of the investigations for pain and/or a mass will show faecal loading of the large bowel.

In older women, diverticular disease should be considered, and many will experience chronic or intermittent lower abdominal pain and distension. The pain tends to be related to defaecation. Occasionally one or more diverticula become acutely inflamed when more severe pain occurs accompanied by fever and signs of localized peritonitis.

5.6.5   Irritable bowel syndrome

Irritable bowel syndrome is a chronic episodic condition which is commoner in women, usually in the third and fourth decades of life. The typical features are of disordered bowel habit, pain, abdominal distension and autonomic disturbances. Since the pain is mostly lower abdominal and in a young woman, the condition is often mistaken for gynaecological disease. Some studies have shown sufferers to be more anxious or neurotic than the general population but this may be a result rather than a cause of the condition as many fear that the symptoms represent serious or life-threatening disease. Scars from previous surgery, hysterectomy, appendicectomy or cholecystectomy usually represent previous attempts at treating her problem.

(a)   History

Pain can occur anywhere in the abdomen but is common in the iliac fossae. It is described as a dull ache or nagging pain with episodes of more severe pain, like colic. Some women notice that the pain seems worse just before a period and is also made worse during and after intercourse. This further confuses the issue as these features are suggestive of a gynæcological problem. During the discussion with the woman it pays to concentrate a little on these areas. In irritable bowel syndrome the pain associated with intercourse usually persists long afterwards. Also the pelvic pain before her period will usually be relieved, albeit temporarily, after she has opened her bowels. Other things that she might associate with the pain are abdominal distension (often described as bloating) and the passing of excessive amounts of flatus. Enquiry about flatus and bowel habit might have to be prompted. Bowel habit is usually altered, the complaint being of constipation, diarrhoea or both. Constipation tends to be worse before periods.

Associated autonomic features include nausea, sweating, palpitations and faintness and may be described as a feeling of panic. A few women also have symptoms of an irritable bladder with frequency and urgency of micturition.

Irritable bowel syndrome: features in the history

The pain
Pain, usually in the iliac fossae
Pain relieved by defaecation
Pain can be worse before periods
Pain during and after sexual intercourse
'Nagging pain' with colicky episodes

Bowel function
Abdominal distension, bloating
Varying bowel habit
Increased passage of flatus
Sensation of incomplete bowel emptying

(b) Clinical signs

The abdomen tends to be rather bloated with tympanic areas indicating gaseous distension. The woman will indicate where she feels pain and sometimes even gentle palpation of these areas causes discomfort. Occasionally parts of the colon can be felt as tender, vague sausage like masses. The pelvic examination is usually normal. The abdominal palpation and pelvic examination can cause discomfort and this pain often persists well after the examination has been completed, presumably due to provoking bowel spasm. Routine laboratory tests, radiological examinations or endoscopic assessments show no abnormalities so making the diagnosis depends very much on a typical history in the absence of physical signs.

(c) Treatment

Treatment starts with a careful explanation of the disorder and strong reassurance that no serious, life-threatening disease is present. Pain is controlled with anticholinergic drugs and the autonomic features, when severe, with a small dose of a beta-adrenergic blocker such as propranolol. Bowel symptoms can be relieved by regular balanced meals and added fibre in the diet. The only surgical procedure which is indicated when the diagnosis is in doubt is laparoscopy, to exclude a gynaecological cause for the pain.

Ureteric calculi produce a typical acute severe colicky pain on one side of the abdomen. The presence of blood in the urine will support the diagnosis which is confirmed by ultrasound examination and intravenous pyelography.

Surgical causes of pelvic pain
Appendicitis
Constipation
Irritable bowel syndrome
Diverticular disease
Ureteric calculi

Medical causes of pelvic pain
Urinary tract infection
Sickle cell crisis
Porphyria

5.6.6   Medical causes

Urinary tract infection, especially in older postmenopausal women, may not present with the typical cystitis, frequency and dysuria but with ill-defined suprapubic pain and backache. Urine cultures will confirm the diagnosis.

Recurrent abdominal pain may occur in women with sickle cell disease. The pain is severe, intermittent and recurrent. It is produced by plugs of sickled cells obstructing blood vessels in the gut or mesentery. The diagnosis will be suspected in an Afro-Caribbean woman; she may have had similar episodes previously and declare her disease and a sickledex test will support the diagnosis. The rare condition of porphyria may present with abdominal pain, especially following ingestion of alcohol or certain drugs such as barbiturates.

5.7   Conclusions

Women with acute pain have an organic lesion which can be diagnosed by history, examination and investigation, and appropriate surgical or medical treatment instituted. In contrast, in as many as half the women presenting with chronic pelvic pain, no diagnosis is made. It is easy to dismiss these women as neurotic and fail to treat them since medical and surgical treatments are rarely effective. However, a sympathetic approach to their problem and careful reassurance that no serious disease is present may in itself produce some relief. In cases of chronic pelvic pain with no known cause it is wise to suspect a 'Hidden Agenda', e.g. undisclosed relevant information. Time, patience and a supportive approach will be required to gain the woman's confidence and so enable her to volunteer her major concerns. In some situations referral to a counsellor or to group psychotherapy will provide ongoing support and help her to come to terms with her problem.

Pelvic pain is a common problem both in gynæcology clinics and in general practice. It certainly can represent one of the most difficult diagnostic challenges in gynæcology and its appropriate management makes demands on all clinical skills.

Learning points

Pain can be gynæcological, surgical or medical in origin
Chronic pain can cause depression
Marital dysharmony may cause or be caused by chronic pain
In almost half the women with chronic pain, no diagnosis is made
Diagnostic laparoscopy is a valuable assessment method
Gynæcological pathology may have fertility implications
Simple, clear information will help dispel fears of cancer
Management is determined by cause and future fertility plans

Further reading

Joshi, U.Y. (1993) Pelvic inflammatory disease. *Hosp. Update*, 19(2), 80–90.
Beard, R.W., Reginald, P.W. and Wadsworth, J. (1988) Clinical features of women with chronic lower abdominal pain and pelvic congestion. *Br. J. Obstet. Gynaecol.*, **95**, 153–61.

6 Sexually transmitted diseases

Caroline Bradbeer

6.1 The principles of sexually transmitted diseases

Sexually transmitted diseases (STDs) are common; about half a million people attend a clinic for STDs each year in the UK. They may affect anyone who is sexually active and are not confined to any particular group in society, although they are more common in large urban populations, because of the relative anonymity and opportunities for partner change. The mean age of female STD clinic attenders is about 24 years, with most being between 15 and 35 years.

There are many myths surrounding STDs such as: they only affect people who are promiscuous or unattractive or dirty. In fact the mean number of sexual partners for the year before attendance at an STD clinic for women is under 1.5 and only slightly higher for men; and the idea – 'I can't have got it from her because she was clean' – is patently untrue. Another myth is that individuals, especially health workers, would know if they are infected – 'she's a nurse so it can't have been her'. This is also wrong; STDs are commonly asymptomatic.

6.1.1 What makes STDs different?

Sexually transmitted diseases are distinct from other medical conditions in several ways. First, multiple infection is common since factors leading to the acquisition of one are frequently common to all. Thus, although in most of medicine the practice is to search for a single unifying diagnosis, in the case of STDs the finding of one should always lead to a search for others.

Secondly, patients cannot be regarded in isolation since, for any episode of STD, there must have been at least three people involved: the index case (i.e. the one presenting first), their contact or contacts and the contact(s)'s contact. It is essential to keep this in mind since, unless everyone involved is treated properly and before they resume sexual relations, reinfection will occur.

The third major difference is that in many cases the greatest morbidity is caused by the psychological, rather than physical, consequences of the disease. Many STDs are easy to treat or relatively harmless; their main effect on the patient is to engender guilt or fear because of the social stigma, the effect on relationships, or the risk of future infertility or cancer.

6.1.2   Fear of having a sexually transmitted disease

A woman may suspect that she has an STD for many reasons: she may have symptoms which she believes are due to an STD or because she knows or suspects a partner is infected. She may want a check-up because she has discovered that her partner has other women, or she may have had another partner herself (a common scenario is the woman who goes on holiday with a group of girlfriends and has a holiday 'fling'). The psychological background can thus be highly charged and the consultation difficult. A small minority become convinced that they have an infection without good grounds and with negative investigations. This venereophobia is uncommon and those who at first sight seem to be in this category, need to be given the benefit of the doubt since there is always the possibility of less than complete disclosure. Both situations are particularly common with the most serious STD of all, human immunodeficiency virus (HIV) infection.

6.2   How do sexually transmitted diseases present and what are the symptoms? (Table 6.1)

6.2.1   Vaginal discharge

The commonest symptom of STDs in women is vaginal discharge, usually accompanied by vulval itching or soreness. The discharge may originate from either the vagina or the cervix, although a discharge arising from the cervix alone can be relatively insignificant and may pass unnoticed, as it has a smaller surface area and is at a greater distance from the introitus. This is important because it explains why the serious infections, *Chlamydia trachomatis* (CT) and gonorrhoea (caused by *Neisseria gonorrhoeae*, also known as the gonococcus and hence usually abbreviated GC), which infect the cervix, cause minimal

Table 6.1 Summary of common sexually transmitted diseases

Disease	Symptoms and Signs	Diagnosis	Treatment	Contact Tracing
Candidiasis	Discharge and itching	Yeast culture from vaginal wall	Antifungals	Not necessary
Bacterial vaginosis (BV)	Offensive discharge	Culture for *Gardnerella vaginalis*	Metronidazole	Desirable
Trichomoniasis (TV)	Watery discharge	Culture for *Trichomonas vaginalis*	Metronidazole	Essential
Herpes genitalis	Viral prodrome	Culture sores for HSV	Acilovir	Desirable
Genital warts	Painless growths, mainly external	Clinical impression	Destruction	Essential
Chlamydia	Asymptomatic or mild discharge. May cause PID	Cervical swab for *Chlamydia trachomatis*	Tetracyclines or erythromycin	Essential
Gonorrhoea	As chlamydia	Cervical swab for *Neisseria gonorrhoeae*	One of many antibiotics, e.g. penicillin	Essential

symptoms. Common infections of the vagina are *Candida* spp – 'thrush', bacterial vaginosis (BV) and *Trichomonas vaginalis* (TV). All these cause discharge as their major symptom but TV is the only one which is usually sexually transmitted. It causes an unpleasant discharge but has no proven serious effects; its importance lies in the fact that its presence is commonly a marker for other STDs, especially gonorrhoea and chlamydia.

The clinical appearance of a discharge can do no more than give a clue to its aetiology and diagnosis rests firmly on microbiological investigations. Discharge is produced by an inflamed vagina, when that inflammation spreads to the drier, more sensitive skin of the vulva it causes itching and, when severe, soreness. *Trichomonas vaginalis*, which tends to cause a profound inflammatory response, is usually associated with a high volume discharge (occasionally bad enough to require sanitary pads) and with vulval soreness rather than itching (Figure 6.1). Candida, on the other hand, causes a less profuse discharge, which may be curdy; the vulval symptom is mainly itching unless the infection is very severe. BV is more in the nature of a bacterial overgrowth of

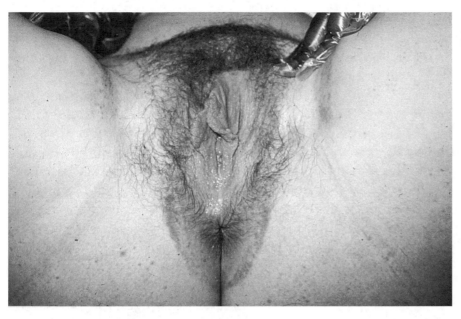

Figure 6.1 Vulval inflammation associated with trichomonas infection. © Anona L. Blackwell. Reproduced with permission.

anaerobes, it causes very little inflammation and minimal vulval symptoms, patients mainly complaining of the fishy smelling discharge, which is caused by the anaerobes breaking down natural amines and releasing ammonia. A fishy smell is often associated with TV infection too, whereas candida has a mild, inoffensive, some say yeasty, smell. The strongest smelling discharge, however, comes from a retained tampon; missing this diagnosis can be very embarrassing.

Recurrent infection with TV is likely to be due to reinfection from a partner, either because he was not adequately treated or because he has himself been reinfected from another woman. Recurrences of BV and candidal infection, however, are commonly spontaneous and usually no cause can be found. Bacterial vaginosis is more common in association with intrauterine contraceptive devices (IUCDs) and other sexually transmitted diseases for which it may be a marker. Candida has been linked to many things ranging from the combined oral contraceptive (COC) to diet and nylon underwear. The only proven predisposing conditions for candidal infection are immunosuppressive conditions, pregnancy, diabetes and broad spectrum antibiotics. Patients with frequent recurrences, however, can become desperate and often employ home remedies such as applying yoghurt intravaginally and eating yeast-free diets; they may also stop taking the COC and become pregnant as a consequence. With these patients it is important to check that they really do have recurrent candida since many other causes of vulval discomfort or discharge can mimic it. Another relevant factor is the association between genital symptoms and psychosexual problems both as a cause and as a manifestation. For example painful intercourse, due to vulvitis, may lead to subsequent avoidance of intercourse or to vaginismus, which itself makes intercourse painful. This can then begin a vicious cycle in which 'thrush' is blamed for chronic symptoms long after the organism has been eradicated. Illustrating the contrast, however, is the woman with a pre-existing psychosexual problem who complains initially of a physical symptom such as discharge, as a more acceptable way of obtaining help. It is important to bear this in mind, especially if symptoms are inconsistent or there is vaginismus.

Physiological discharge occurs during times of hormonal flux. Puberty, pregnancy and the menopause are the commonest causes but ovulation must also be considered. Patients are generally well aware of their cyclical changes and know what is normal for them. In certain situations, however, this may not be the case. One of these occurs when a woman stops taking the COC after several years. Having forgotten what a normal cycle was like, she views the reappearance of her ovulatory discharge as abnormal. Another common scenario is the girl with late menarche whose is already aware of the symptoms of genital infections and becomes afraid of what her new, physiological, secretions may signify.

Cause	Type of discharge	Other symptoms
Physiological	Clear or white and inoffensive	None usually
*Neisseria gonorrhoeae**	Very light if any	Can cause pelvic inflammatory disease
*Chlamydia trachomatis**	Very light if any	Can cause pelvic inflammatory disease
Retained Tampon	Heavy and malodorous	
Bacterial vaginosis	Fishy smelling vaginal discharge	Minimal symptoms, if any
*Trichomonas vaginalis**	Profuse, green, 'frothy' and offensive	Vulval soreness
Candida	Curdy, white and inoffensive	Vulval itching

*Indicates sexually transmitted

6.2.2 Vulval ulcers

Vulval soreness is common in the vaginal infections mentioned above and there may even be some fissures in cases of severe candida. Frank, demarcated sores, however, are not seen in these conditions. The commonest cause of such sores is herpes genitalis caused by *Herpes simplex* virus. (Both types 1 and 2 can cause genital infection although type 1 is usually associated with the common cold sore; Figure 6.2.) Other causes are trauma, which may be self-inflicted or due to overvigorous intercourse or rape; systemic disorders such as Behçet's and Crohn's diseases and, in later life, malignancies. Syphilis causes painless ulcers and is rare in this country but should always be considered because of its serious sequelae if left untreated; it is still a likely diagnosis in women who themselves, or whose partners, have been abroad.

Causes of vulval ulcers

Herpes genitalis
Trauma
Behçet's disease
Crohn's disease
Syphilis

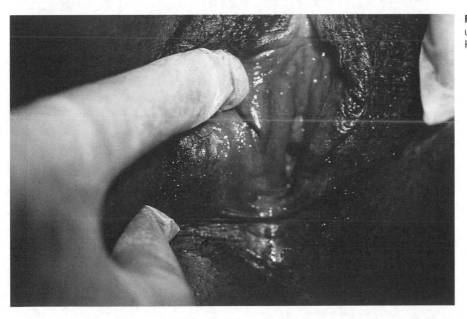

Figure 6.2 Herpetic vulval ulceration. © Anona L. Blackwell. Reproduced with permission.

(a) Herpes genitalis

This is a recurrent condition, the first attack represents a new infection but thereafter recurrences occur in about 50% of sufferers without further exposure, often apparently triggered by hormonal changes or stress. Classically the patient complains of vulval sores preceded by viraemic symptoms known as a prodrome; the sores can be severe enough to make walking difficult and inhibit micturition. Indeed, before treatment became available, patients were often admitted for suprapubic catheterization.

Herpes sores begin as erythematous areas on the vulva bearing small blisters which burst to form shallow ulcers and then crust over; the history can be almost diagnostic, especially if blisters have been seen. This whole cycle, from the prodrome to complete healing may take three weeks in a first, untreated, attack, although subsequent recurrences generally heal within seven days. Some people do get severe recurrences, occurring perhaps monthly for years, but this is rare. The more usual pattern is of one or two attacks in the first year with declining severity and frequency for the next two or three years before they cease altogether. However, women who suffer mild cases may never be diagnosed, or may present for the first time with one or two genital ulcers with or without a history suggestive of recurrent attacks, often previously labelled as 'thrush'.

Herpes has had a bad press. Patients know that it is 'incurable' and believe that it will prevent them from having a normal sex life and from bearing children. A diagnosis of herpes thus has massive psychological overtones. It is common for a patient, on receiving a diagnosis of herpes after years of mild recurrent disease, to break down over what is really just a new label for an old condition. Usually, infection can only be transmitted or detected when sores are present. A woman can, therefore, acquire herpes genitalis within a monogamous relationship from her partner if he has a recurrence of pre-existing disease. If she has herpes, she is unlikely to transmit it to a new partner provided she avoids intercourse during times when she has symptoms. Asymptomatic viral shedding does occur but for practical purposes, can usually be disregarded.

Herpes simplex infection of the newborn is usually a consequence of transmission from sores in the mother's genital tract during delivery. However, it is very rare and affected mothers frequently give no history of herpes. Women with herpes can therefore be reassured that any risk to a baby would be minimal and only likely at all if there are active herpetic sores during vaginal delivery. Delivery by caesarean section should be considered if active herpes is present at the time the membranes rupture in labour, although if the membranes have been ruptured for several hours (usually more than 6 hours) little benefit will accrue.

6.2.3 Dysuria

It is useful to divide dysuria into true dysuria, originating from the bladder, and vulval dysuria where the pain is a result of urine passing over a sore, inflamed vulva. The causes are very different, the former being related to the urinary tract, usually a bacterial infection, and the latter to the genital tract. The two can usually be distinguished just by examining the vulva and urethra where localized erythema or ulceration may be obvious. If the cause is a bladder problem, the external genitalia will look normal.

With a little explanation patients are usually able to make the distinction and this can save inappropriate investigations and treatment. A typical example is the woman with vulval dysuria who goes to her general practitioner complaining of 'cystitis'. If her symptoms are not clarified she may receive antibiotics for a presumed urinary tract infection, whereby her symptoms, actually due to candidal vulvitis, will be worsened.

6.2.4 Lumps and bumps

As in much of medical practice, a patient who finds she has a 'lump' will present relatively quickly. Most of these lumps are genital warts. Others, however, are benign cysts, furuncles or, commonly, normal anatomy. This last is usually the hymenal tags which the patient has found through self-examination. It is important to reassure her that all is normal but also to find out what has prompted the examination since it is a difficult thing to do, requiring a complicated arrangement of mirror and light, and may have been precipitated by some genital symptom or fear of STD or cancer.

(a) Genital warts

Patients with genital warts usually present giving this diagnosis since they look like warts elsewhere on the body (Figure 6.3). They are mainly found on the external genitalia but can occur anywhere in the genital tract and perianal area; the latter is a common site and should always be examined. Warts are painless but may itch, especially if they

Figure 6.3 Genital warts. ©
Anona L. Blackwell. Reproduced
with permission.

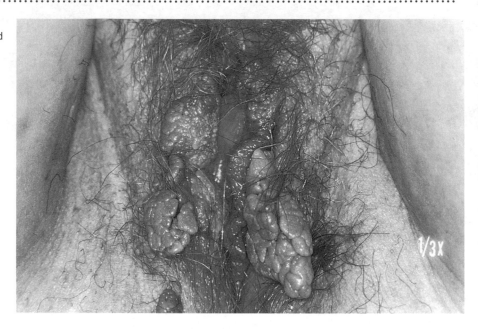

are large. This is probably due to organisms, such as candida, trapped
in the crevices of the warts setting up a mild inflammatory reaction.
Genital warts are of some concern because of their links with cervical
cancer. In fact only certain subtypes of human papillomavirus (HPV 16
and HPV 18 particularly) are associated with cervical cancer. The types
seen in association with obvious warts are benign papillomaviruses
such as HPV6. Patients, however, know about the link and worry.
Another problem with genital warts is their incubation period which
can be several months. This means that they can arise, like herpes, *de
novo* during a long-term stable relationship causing understandable
friction and mistrust. Treatment is aimed at destroying the warts,
using local antimitotic agents or physical destruction. The viral infection tends to be widely spread through the skin and, perhaps not
surprisingly, treatment is frequently unsuccessful; even if the original
lesions are destroyed more can reappear at other sites and require
further intervention. Rarely, the problem persists for months or years
causing considerable morbidity.

6.2.5   Pelvic pain and cervical infections

Gonorrhoea (GC) and chlamydial (CT) infections can be considered
together for the most part, indeed they are often found as a dual
infection. Both infect the cervix, and to a lesser extent, the urethra and

rectum; both are always sexually transmitted; both cause minimal symptoms in women and urethral discharge in men, and both have as their main complication in women pelvic inflammatory disease (PID).

Chlamydia trachomatis is a more common pathogen than GC in the UK, and seems to produce a milder inflammatory response and thus milder symptoms – which may delay the diagnosis. As an aetiological agent of PID it is thought to account for up to half of all cases, whereas GC is probably responsible for about a sixth. Since most women with uncomplicated GC or CT are asymptomatic, they usually present only if they have symptoms of accompanying STDs, such as TV, or because they suspect a partner is infected. Often the diagnosis is not made until complications, such as PID, infection of Bartholin's glands or disseminated infection, occur. Rarely a mother is diagnosed because she transmits her infection to her newborn baby as it passes through the birth canal. Neonatal infection frequently presents with a discharge from the eye (ophthalmia neonatorum).

Delay in detecting both CT and/or GC may result in tubal infertility. It has been shown that the approximate risk of infertility following one episode of chlamydial PID is 12%, with 35% after two episodes and 75% for three or more.

PID is difficult to diagnose; the symptoms vary from an acute abdomen, through mild pelvic pain, to an asymptomatic form which is only discovered at laparoscopy performed for tubal infertility (Chapter 7). Even when diagnosed, many cases are mismanaged with inappropriate antibiotics or failure to prevent reinfection by contact tracing and treating the partner[s].

Symptoms of mild PID have often been present for weeks before presentation; they are vague lower abdominal pain, often described as resembling dysmenorrhoea, slight vaginal discharge, and deep dyspareunia (this latter being equivalent to tenderness on bimanual examination). Often no organism is found but on examination of the male partner there is a urethral discharge (non-specific urethritis, NSU) and this, even in the absence of an organism, has been shown to recur after treatment if the female partner is not treated. NSU has also been linked to PID.

Once more, the psychological effect may outweigh the immediate physical effect of the illness. Because of the high incidence of asymptomatic infection, it is rarely possible to say how long a woman has been infected (although this can often be inferred from the sexual history). The possibility that acquisition of the infection predated the relationship should be considered, both for the patient's sake and for contact tracing. The patient's concerns about the implications of having an STD are added to by fear of future infertility if she also has pelvic inflammatory disease. Reassurance is difficult since the fears are real and the exhortation to 'try to get pregnant soon' (to try and show her that her fertility has not been compromised) may be unwelcome if she has just discovered that her partner has been unfaithful.

6.3   How to investigate STDs

Investigation consists of history, examination and special tests. This section deals only with those parts which are specific to the investigation of STDs and assumes that a full history and gynaecological examination will be undertaken.

6.3.1   Sexual history taking

Part of clinical training involves being taught to ask in detail about personal things such as bowel habit, but a patient's sex life is rarely mentioned. The art of taking a sexual history is to avoid embarrassment: be factual and never surprised. One way to start is to ask the patient 'When did you last have sex?'. This will give an idea about incubation periods, risk of pregnancy and whether she has a current relationship.

The next thing is to find out whether this is a long-standing relationship, a new one or a casual liaison. Useful questions are 'Was it a regular partner?' (usually a boyfriend or spouse but occasionally another woman) and 'How long have you been together?'. It is then important to establish if there are any other partners so: 'When did you last have sex with someone else?', and, then to ask the same questions about that relationship. It may be necessary to go back several months or years, depending on the duration and nature of the problem. Prostitutes often do not declare themselves as such, but women with a large number of partners and high condom use may admit to being 'a working girl' or 'on the game'.

Questions about contraception come next, with particular attention to use of condoms and unprotected intercourse. It is surprising how many women use no contraception yet freely admit that they do not wish to become pregnant. Finally it is essential to ask about the partner's symptoms. Many women do not volunteer that he has symptoms or even that he has been to a clinic. It is important to know this since it may alter her investigation and/or management. Taking a sexual history is dealt with in further detail in Chapter 14.

6.3.2 Specimen taking

All patients in whom an STD is suspected should be fully investigated, this includes a cervical smear and blood test for syphilis. Specific tests, such as herpes culture, are indicated if sores are present, and HIV antibodies should only be measured after counselling.

Failure to diagnose an STD is usually a result of incorrect specimen taking: the wrong site, the wrong medium or delay in getting the specimen to the laboratory. Sexually transmitted organisms are fastidious by nature, if they were hardier it would be possible to transmit them in less intimate ways. Specimens taken under less than ideal conditions often fail to grow, leading to false negatives. Table 6.1 gives a summary of the sites and specimens necessary but extra explanation is needed for the all-important cervical infections.

GC can be found on a high vaginal swab in only about 50% of cases because it resides within pus cells in the cervix and so is only a 'contaminant' in the vagina. The best specimen for GC isolation is therefore cervical secretions; urethral and rectal swabs may also grow GC and should ideally be taken as well. A throat swab is also indicated if oral sex has taken place especially if this may have been the only form of sexual activity, e.g. in prostitutes. If possible a selective medium should be directly inoculated with the specimen to improve the chances of isolating the organism.

Chlamydia also inhabits the cervix but it is found in epithelial cells so the cervical swab for CT needs to be taken fairly vigorously in order to dislodge enough of these cells. *Chlamydia trachomatis* is a bacterium without a cell wall which makes culture on normal media impossible. Cell culture, as for a virus, is available, together with several other antigen detection methods.

6.3.3 Treatment

The principles of treatment are to use appropriate therapy for an adequate length of time, to test for cure and contact trace to prevent reinfection. Therapy for common STDs is outlined in Table 6.1 and details will be found in textbooks of genitourinary medicine. The rationale for tests of cure is based on the asymptomatic nature of so many STDs which means one cannot rely on the disappearance of symptoms to imply cure.

Treatment sometimes cannot wait for a microbiological diagnosis and may have to be initiated on a 'best guess' basis. This is especially so in the case of PID; here an antibiotic must be chosen to cover all likely organisms. Since CT has no cell wall, drugs such as penicillins and cephalosporins are not appropriate and the drug of choice is a tetracycline or, if there is a risk of pregnancy, erythromycin.

Contact tracing requires immense tact, patience and much explanation. The mainstay is confidentiality. There is an Act of Parliament which forbids disclosure about an STD (including HIV) to a third party, except to the medical practitioner, and only then if it is essential for medical care. Therefore, a patient can be reassured that, if she brings a partner to the clinic, he will not be told anything about her sexual history or diagnosis. Contact tracing can still be difficult. Contacts may be untraceable; 'I was drunk – I think he was called Harry'. They may refuse to attend: 'He says there's nothing wrong with him'. Couples may play ping-pong with their infection: 'Oh, yes, we had sex this morning – I know I haven't been treated yet but he'd finished his tablets'. Patients (usually the men) with multiple contacts may not tell all of them: 'I know she only has sex with me so there's no need for her

---

Learning points

STDs are often asymptomatic, especially in women

STDs are frequently present as multiple infections with different organisms

Sexually transmitted diseases are common and diagnosis rests heavily on appropriate investigation

Sexual history taking is an important clinical skill and an integral part of the assessment of patients suspected of having contracted a sexually transmitted disease

Contact tracing is an essential part of the overall management to prevent recurrence of infection

Most sexually transmitted diseases are easily treated

The psychological background and stigma attached to these conditions is of equal import and demands sensitivity and professionalism from health attendants

to be seen'. There may be just plain denial; a woman said to me once, 'I don't know how gonorrhoea got into our family'.

Further reading

Alder, M.W. (ed.) (1990) *ABC of Sexually Transmitted Diseases*. British Medical Journal, London.

Alder, M.W. (ed.) (1993) *ABC of AIDS*. British Medical Journal, London.

Miller, D.E. (1992) *The Management of AIDS Patients*. Macmillan, London.

Wisden, A. (1989) *Colour Atlas of Sexually Transmitted Diseases*. Wolfe, London.

7 Infertility and pregnancy loss

Susan Blunt and David Walker

7.1 Introduction

The majority of human beings have a strong desire to reproduce and take it for granted that conception will quickly follow unprotected intercourse and that a healthy baby will follow conception. In reality the converse is true – it is far more likely that conception will not follow unprotected intercourse (fecundity rates are as low as 33% in the first month of trying, i.e. 67% will not conceive, and thereafter decline to approximately 5–10% each month). A live birth occurs in only half of all pregnancies – one fifth of clinical pregnancies are terminated in England and Wales and just under a third miscarry after implantation. Compared to many mammals, e.g. dogs and horses etc., our fertility is dismally poor and the human sperm count is among the lowest found in all primates. Infertility and miscarriage are all part of the spectrum of reproductive failure and as the above statistics show, is one of the commonest reasons for gynaecological referral. For the couple concerned, reproductive failure, causes a great deal of grief and anguish – principally for the woman. There is a sense of failure, lack of self-esteem, guilt and isolation from their cohort of friends who have children. Reproductive failure is therefore both a psychological and social as well as a medical problem. In addition, infertility investigations probe into the most intimate and private area of a couple's life, as the history involves questions about the couple's sex life and the examination commonly incorporates a genital examination. Couples should therefore be treated promptly, compassionately and given an accurate and realistic prognosis for their achieving parenthood.

7.2 Infertility

7.2.1 Incidence of infertility

About one in six couples seek specialist help at some time in their lives because of their inability to conceive. Unfortunately half of these will remain childless or fail to achieve their desired family size. Although, overall, the incidence is not increasing infertile couples are seeking help more commonly these days and have higher expectations following widespread media coverage of so-called 'miracle births'. At the end of one year of trying 10% of fertile couples will not be pregnant and 5% will still not be pregnant after 2 years. Therefore, infertility can be defined as 'difficulty in achieving a pregnancy for more than 12–24 months'. When the woman is less than 30 and there is no obvious cause for the couple's infertility in their history, then investigations should be deferred for 24 months. However, when the woman is over 35 or when there is an obvious cause in the history, then investigations should be started sooner. Bearing the natural conception rates in mind, when no abnormality is found and couples have been infertile for less than 3 years they should be reassured. Similarly when an effective treatment is started, it should be given a fair chance and couples fully acquainted with the conception rate expected so that they do not anticipate an instant conception.

7.2.2 Causes of infertility

There are many potential causes. The following are essential for conception to occur.

1. There must be oocytes within the ovary which must mature, undergo meiosis and release
2. The fallopian tubes must pick up the released oocytes and transport them actively towards the uterus.
3. As fertilisation occurs in the tube by ascending sperm, the uterus, cervix and vagina must be structurally and functionally normal.
4. Normal numbers of actively motile sperm must be deposited high in the vagina so that they can readily penetrate normal, non-inflammatory, non-hostile cervical mucus.
5. The fertilized embryo must have the ability to implant within the uterine lining and the uterus must be receptive.

Figure 7.1 Causes of primary and secondary infertility. Note the increased frequency of tubal causes in secondary infertility.

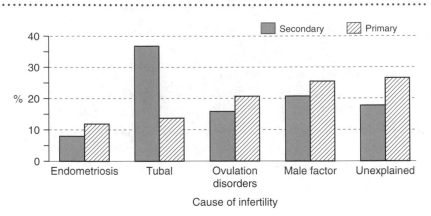

Figure 7.1 shows the frequency with which the main causes for infertility occur in patients with primary infertility (those who have never conceived before) and those with secondary infertility (a confirmed previous pregnancy).

7.2.3 Investigation

There may be single or multiple causes of infertility and therefore all the causes should be investigated so that the appropriate treatment or treatments can be given.

(a) History and examination

Ideally the couple should be seen together in a relaxed, friendly and private environment. The doctor should adopt a positive but realistic attitude. The couple should be asked about the duration of infertility and any past history of operations and illnesses, especially past pelvic surgery, sexually transmitted diseases, pelvic inflammatory disease and endocrine disorders. Previous fertility or infertility with this or another partner is relevant, plus the outcome and complications of any pregnancies. Asking about previous pregnancies might be a sensitive issue particularly if the woman has had a previous termination. This will naturally be a prominent feature on the 'hidden agenda' and may be a source of guilt feelings.

The couple should be asked about current medication, occupation and smoking habits. Details of coital frequency or coital problems should be sought as tactfully as possible. Details of the woman's menstrual history are important as this may indicate disorders of ovulation.

The couple should be examined generally and specifically. The examination should include the body mass index of the woman – as being too fat or too thin (as in anorexia nervosa) can impair ovulation. Hirsutism and galactorrhoea in the woman may also indicate an endocrine cause for infertility. Genital examination should detect any deviation from normal that may or may not be significant, e.g. in the male, hypospadias may prevent the sperms being ejaculated high enough in the vagina to penetrate cervical mucus. In the woman, uterine fibroids, although an obvious abnormality, only rarely cause infertility. It is debatable whether a varicocoele contributes to male subfertility.

Subjects to cover in the history

Duration of the problem
Previous pregnancies and their outcomes
How often intercourse takes place
Any problems with intercourse
Any problems with menstruation
Previous contraception methods
Previous infections, especially pelvic inflammatory disease
Previous pelvic surgery

(b)   Investigation of the woman

Ovulation   Progesterone is only found in the postovulatory or pregnant woman, and therefore directly or indirectly detecting progesterone is the basis for most tests of ovulation. Serum levels of progesterone taken in the mid-luteal phase, i.e. one week before the next period, should be at least greater than 30 nmol/l. A basal body temperature chart (usually recorded daily, first thing in the morning) reflects the thermogenic action of progesterone on the hypothalamus; a rise in basal body temperature indicates ovulation. However, the temperature chart has many drawbacks – the temperature can shift with suboptimal levels of progesterone and in some ovulatory women the temperature does not change with ovulation. There is a danger that some women become obsessed by their charts and others become upset by the daily reminder of their infertility. Their use should therefore be limited.

Ultrasound is a very accurate method of detecting ovulation but it is very time consuming as frequent visits to the ultrasound department are required to serially assess the growth of a follicle and its eventual rupture. Urinary luteinizing hormone (LH) kits are promoted in women's magazines and therefore are commonly used by infertile women. They detect the ovulatory surge of LH, are accurate but expensive. About 80% of women with oligomenorrhoea have polycystic ovarian syndrome (PCOS) characterized by anovulation and elevated serum LH levels. Obesity and hirsutism are common and there is usually elevation of the serum LH (>10 iu/l), LH to follicle stimulating hormone (FSH) ratio (>2.5) and androgens (testosterone and androstenedione).

The main causes of amenorrhoea are PCOS, hyperprolactinaemia, hypogonadotrophic gonadism and premature ovarian failure. A serum LH, FSH and prolactin test will differentiate between these causes. These main causes of amenorrhoea are discussed in Chapter 4.

Tubal patency   A diagnostic laparoscopy with hydrotubation using methylene blue dye is an excellent and comprehensive way of accurately assessing the woman's internal genitalia (Figure 7.2). It is important to explain to the couple that this is a diagnostic procedure. It is a minor operative procedure performed under a general anaesthetic, usually as a day case. It does provide a great deal of valuable information that is often vital if treatment is to be properly planned. The extent of any tubal damage and occlusion can be accurately assessed together with peritubal and periovarian adhesions. Pathology such as endometriosis can be seen and often treated. In addition ovula-

Figure 7.2 Methylene blue from patent fallopian tubes at laparoscopy.

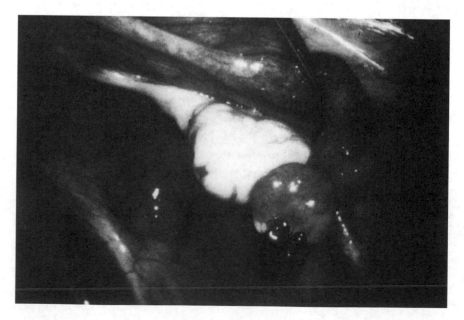

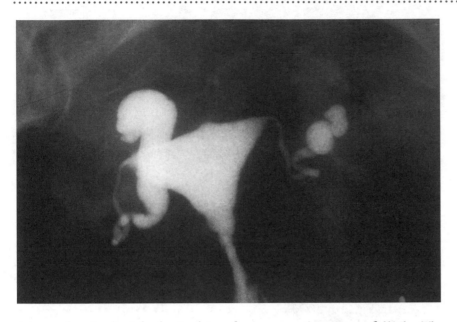

Figure 7.3 Hysterosalpingogram demonstrating bilateral hydrosalpinges.

tion can be checked for by looking for a corpus luteum or follicle. The procedure is usually combined with a hysteroscopy (endoscopic examination of the uterine cavity) and endometrial biopsy, to assess the uterine cavity and to look for secretory change in the endometrium histologically. Secretory endometrium usually correlates with the luteal phase in the ovary and indirectly indicates that ovulation has taken place. Occasionally endometrial TB culture is of value. Hysterosalpingography (radiological contrast imaging of the uterine cavity and fallopian tubes) is reserved for women in whom a laparoscopy is contraindicated and is important prior to planned tubal surgery to define the level and extent of tubal damage (Figure 7.3).

(c) Investigation of the man

Standard investigation for male infertility is analysis of seminal fluid. Seminal fluid samples are obtained by masturbation, a procedure that some men find difficult to do on demand and that others, a minority, refuse to do. Sperm counts vary quite considerably from day to day, it is important to check a minimum of two samples at least a month apart, especially if the first sample is suboptimal in any of its characteristics (see below).

The World Health Organization criteria for a normal seminal fluid analysis are given in Table 7.1.

The postcoital test (PCT) assesses sperm–mucus interaction. A sample of mucus is collected from the woman shortly after she has had intercourse. Microscopic examination should show normal forward

Table 7.1 WHO criteria for normal seminal fluid

Volume	>2 ml
Sperm concentration	>20 × 10⁶ sperm/ml
Total sperm count	>40 × 10⁶ sperm
Motility	>50% with forward progression
Morphology	>50% normal morphology
Antibody coating of the sperm (MAR test)	<10% of the sperms affected

motion of the sperm through the mucus. Correct timing is crucial as the woman produces good quality mucus only in the 24–48 hour preovulatory phase of the cycle; the commonest reason for a poor result is bad timing. This is where temperature charts or ovulation predictor kits can help in planning when to do the test.

7.2.4 Treatment

(a) *Treatments of ovulation failure*

With the exception of premature ovarian failure, all anovulatory patients can be successfully treated to achieve a normal cumulative conception rate. First, any abnormality of thyroid function or weight should be corrected. Women with hyperprolactinaemia should first have investigations to exclude a pituitary tumour following which they can be started on bromocriptine or cabergoline. Bromocriptine can cause nausea and should initially be started in a low dose, gradually building up over 4 weeks to 5 mg/day.

All other endocrinopathies should be treated initially with clomiphene (50–100 mg from day 2 to 6 of the cycle). If ovulation fails to occur after 3 months, then human chorionic gonadotrophin (HCG) should be given if ultrasound shows there is a follicle developing but not rupturing, i.e. the LH surge is either absent or deficient. If this fails or if the woman does not conceive after 9 months of clomiphene treatment then gonadotrophin treatment should be considered. This is a complex, expensive therapy prone to serious complications such as multiple pregnancy (1 in 4 risk) and ovarian hyperstimulation (Chapter 3). Therefore this treatment requires close ultrasound and endocrine monitoring. About 10% of clomiphene failures are due to hypothalamic failure and these women have an excellent response to pulsatile luteinizing hormone releasing hormone (LHRH). This type of treatment requires a special pump that will deliver the hormone in a pulsatile fashion, mimicking the release from the hypothalamus. This type of treatment is very safe and effective in properly selected women.

(b)   Sperm dysfunction

Unfortunately the vast majority of sperm dysfunctions do not have a correctable cause. If the man is azoospermic or has very severe oligospermia ($<5 \times 10^6$ sperms/ml) then there may be either an obstructive element (blocked vas deferens) or an endocrine cause. The man's pituitary and gonadal hormones should be checked. Low levels indicate hypothalamic or pituitary dysfunction that can be treated with gonadotrophins etc. as in the woman; elevated gonadotrophin levels indicate untreatable testicular failure and normal levels may suggest an obstructive cause which can be investigated by a testicular biopsy and a vasogram.

If no endocrine or obstructive cause for the oligospermia can be found then no treatments have been shown to be of proven benefit. In particular it is worth noting that artificial insemination using the husband's sperm (AIH) and/or hormonal therapy do not work. Men with antisperm antibodies may benefit from high dose steroids but this treatment is risky and can only be given for a few cycles. Donor insemination, although easy, simple and effective treatment, presents the recipient couple and the donor with complex ethical considerations. Frozen semen is always used so that the donor can be screened for infections including human immunodeficiency virus. Assisted conception (section 7.3) can be of benefit and in particular *in vitro* fertilization (IVF) can be valuable for testing the fertilizing ability of spermatozoa and treating the problem. However, the pregnancy rates are lower in couples with sperm dysfunction.

(c)   Tubal disease

Fallopian tubes are not passive drainpipes into which the oocyte falls and rolls down into the uterus. The fimbriae actively scan over the surface of the ovary at ovulation time and gently 'suck' the oocyte into the tube where cilia waft it into the uterus. Fertilization occurs during transit and tubal secretions nourish the early embryo. Tubal disease commonly permanently damages and destroys this cilial lining so that even though tubal patency can often be restored by surgery, the cumulative conception rate is poor (12% for severe disease to 68% for mild disease after 2 years). Surgery is therefore best for mild to moderate tubal pathology with IVF being increasingly used as primary treatment for severe tubal pathology. If conception has not occurred one year after surgery, then IVF is the next treatment option.

(d)   Endometriosis

It is debatable whether minor degrees of endometriosis cause infertility. If it does then none of the currently available treatments improve fertility rates. Moderate and severe disease does cause adhesions and ovarian cysts. Both of these can impair fertility. Surgical treatment is often necessary to restore normal anatomy and this is an area where laparoscopic laser surgery has shown encouraging early results.

Medical therapy, usually danazol or LHRH analogue therapy, also has a place in the management of moderate and severe disease. However, if conception has not followed within one to two years after completion of anti-endometriosis therapy, then assisted conception should be considered.

(e)   Unexplained infertility

This term describes the situation where no known cause for failing to conceive has been found. It is always unsatisfactory in clinical medicine to fail to find a cause for a patient's suffering and this is certainly the case with infertility. If the couple have been trying for less than 3 years they should be reassured on the basis of the normal cumulative conception rates. After 3 years fecundity rates fall off markedly and so assisted conception becomes the treatment of choice.

7.3   Assisted conception

Assisted conception is the term used to describe IVF, gamete intrafallopian tube transfer (GIFT), zygote intrafallopian tube transfer (ZIFT), intrauterine insemination (IUI) and numerous other variations on a theme. IVF was the first successful assisted conception technique resulting in the birth of Louise Brown in 1978. IVF today is the most widely used assisted conception method and is very successful giving a nearly normal cumulative conception rate. The main indications for IVF are tubal damage, unexplained infertility, endometriosis and male subfertility. IVF involves stimulation of multiple ovarian follicles from which mature oocytes are aspirated, usually under ultrasound control,

Table 7.2 IVF treatment cycle summary

1. Pituitary suppression with LHRH analogue (14+ days)
2. Ovarian stimulation with human menopausal gonadotrophins (10–14 days)
3. Administration of HCG
4. Transvaginal oocyte collection under ultrasound control
5. Addition of prepared sperm
6. Confirmation of fertilization
7. Replacement of embryos into uterus after 48h in culture
8. Luteal phase support

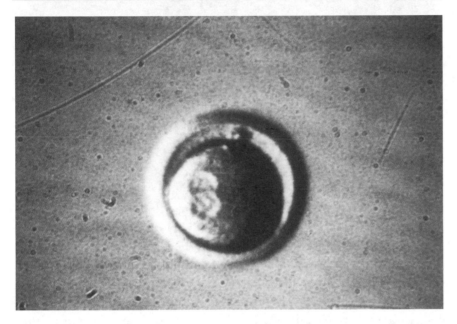

Figure 7.4 Human egg with two pronuclei 18h postfertilization.

and fertilized outside the woman's body (Table 7.1). Up to three of the resulting embryos (Figure 7.4) are transferred back to the uterus transcervically about 2 days later. In cases of very severe sperm dysfunction, micromanipulation and injection of a single spermatozoon into the oocyte has been attempted with some success (Figure 7.5). Lower success rates are seen when the woman is over 40 or when poor quality sperms are used. Donor eggs and donor sperms can be used if the couple's own gametes are either absent or defective.

7.4   Counselling

Although half the patients treated can be helped to achieve their aim of having a child it may cost the couple dearly both emotionally and

Figure 7.5 Methods of micromanipulation.

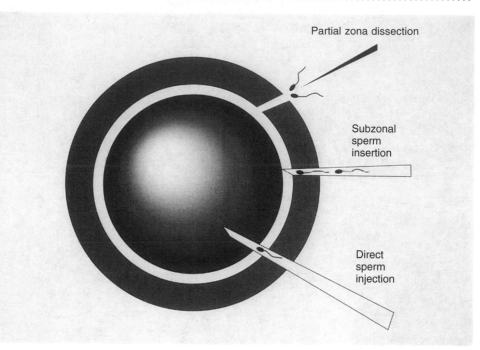

Partial zona dissection

Subzonal sperm insertion

Direct sperm injection

financially because regrettably IVF is mainly only freely available within the private sector. As mentioned earlier, infertility itself together with associated investigations, are very stressful and the treatments often add to the stress. There is therefore a high incidence of depression, marital breakdown and sexual dysfunction. Counselling can help couples overcome their feelings of failure and lack of self-esteem which facing up to a childless life may bring. Indeed a trained counsellor is now an obligatory part of an assisted conception team.

7.5   Adoption

Unfortunately with the ready availability of terminations combined with society's greater acceptance of single parents, there are very few babies now available for adoption. Although adoption does offer a couple the prospect of achieving their desired goal of a family it too can be a difficult and rigorous process. The various and many interviews and formalities that take place are seen as yet more barriers that the couple must overcome. Most couples seeking adoption will have already experienced the investigative and treatment phases of infertility management and will require continued support and encouragement.

Adoption should always be mentioned early on in the counselling process as one of the options of 'Management'. Introducing it at the end of a protracted and failed treatment is not satisfactory and may compound further the sense of failure and loss that is so often seen in couples who have reached this stage. Most agencies that manage the adoption process will have an age limit for accepting prospective couples who generally must not be receiving active fertility treatment at the time. Forward planning is important, particularly in older couples.

7.6 Pregnancy loss

Although a woman may understandably equate her positive pregnancy test with the certain birth of a baby, the reality is different. Preclinical pregnancy loss rates (i.e. before 6 weeks of pregnancy) have been found to be as high as 22% and a further 12% of pregnancies beyond 6 weeks miscarry. The term 'abortion' should be avoided when talking to couples who have had miscarriages, as the term is commonly taken to mean an induced or therapeutic abortion. Miscarriage is, therefore, defined as unplanned loss of a pregnancy after the missed period and up to the 24th week of gestation. In most clinical situations no cause can be found to explain why a woman has miscarried. This should be explained to the couple, emphasizing that although a tragic personal loss it does not mean that they are in any way abnormal and in particular, the miscarriage was not the result of anything that she or they did. Women often blame themselves and a common question is 'Was it something that I did?'

7.6.1 Recognized reasons for miscarriage

1. **Chromosomal defect** which is found in a third of all abortuses. These are commonly trisomy, monosomy, triploidy, or translocation.
2. **Genetic defects**
3. **Hydatidiform mole**
4. **Hormonal** – high serum LH levels in the late follicular phase of the cycle cause abnormal maturation of the oocyte and high miscarriage rates.

5. **Uterine abnormalities** (fibroids, bicornuate uterus etc).
6. **Cervical incompetence** – this commonly causes miscarriage in the mid-trimester.
7. **Infections** such as listeria, toxoplasmosis, tuberculosis, syphilis, rubella and cytomegalovirus.
8. **Acute illnesses/pyrexia** such as pyelonephritis, appendicitis etc.
9. **Immunological causes**

One in 250 pregnancies is ectopic and ectopic gestation is a relatively more common conclusion to a pregnancy in infertile women. This is an upsetting form of pregnancy loss because not only does the woman lose a pregnancy but the condition causes significant maternal morbidity and even death. It often involves the loss of the fallopian tube and a laparotomy. Fortunately small, unruptured ectopics can now be dealt with laparoscopically with conservation of the fallopian tube.

A hydatidiform mole is a benign trophoblastic tumour but may have malignant potential, progressing to a choriocarcinoma. A hydatidiform mole is a rare form of miscarriage in the United Kingdom occurring once in approximately 1000 births. It is commoner in women over 40. Usually there is no fetus present and this is termed a complete hydatidiform mole. Less commonly, triploid fetal material can be identified in association with a trophoblastic tumour, when the condition is known as a partial hydatidiform mole.

7.6.2   Diagnosis and differential diagnosis

Vaginal bleeding with a viable fetus and closed cervical os is called a threatened miscarriage. The outlook is good and no treatment is of proven benefit. The presence of low abdominal 'cramp-like' pain usually indicates that the miscarriage is becoming inevitable which is diagnosed once the internal cervical os is open. If the products of conception are partially expelled then the miscarriage is incomplete. If the products of conception are stuck in the cervical os a marked vagal response may be provoked causing so-called cervical shock, with the patient becoming collapsed and hypotensive (in addition to her anxiety and severe psychological stress). The products of conception should be removed from the cervix immediately. If the products of conception are passed intact which commonly happens before 6 and after 12 weeks gestation, the os closes, the pain and bleeding stops and this is called a complete miscarriage. If, however, the fetus dies but is retained within the uterus, this is a missed abortion. Usually there is little or no bleeding. There is the danger of a disseminated intravascular coagulopathy developing if the products of conception are septic or retained for greater than 4 weeks. Occasionally only an empty gestational sac can be identified (by ultrasound scan): this is called a blighted ovum.

Hydatidiform moles are diagnosed by a typical 'snow storm' appearance on scan. The uterus is usually large for dates, the symptoms of pregnancy such as nausea are often exaggerated and Doppler ultrasound examination fails to find a fetal heart. Occasionally the woman passes hydropic vesicles. Although a ruptured ectopic pregnancy is easy to diagnose – the patient is collapsed with peritonism – an unruptured ectopic pregnancy is notoriously difficult (Chapter 3).

7.6.3 Investigations and management

The investigations that are most often helpful in the differential diagnosis are full blood count, pregnancy test and an ultrasound scan. All patients should have their blood grouped and saved so that if they are Rhesus negative, anti-D can be given within 48 h of pregnancy loss to minimize the risk of subsequent isoimmunization.

A threatened miscarriage requires no specific treatment and can be managed at home. Both an inevitable and incomplete miscarriage should be treated by evacuation of the uterus under general anaesthesia to prevent life-threatening bleeding and sepsis. The products of conception should always be sent for histological examination. If there is any clinical suspicion of an ectopic pregnancy a laparoscopy should be performed because it is much safer to subject the woman to a laparoscopy than to miss an ectopic pregnancy. Patients with a missed abortion should have a full coagulation screen followed by evacuation of the uterus – either surgically if the uterus is less than 14 weeks, or medically using vaginal prostaglandins if it is larger than this. In patients with a hydatidiform mole the uterus must be emptied by suction evacuation as this lowers the incidence of patients requiring chemotherapy. All women in the UK must be followed-up at either Charing Cross Hospital, London, Dundee or Sheffield (these are Supra regional Centres for HCG assay and treatment of choriocarcinoma) for at least six months and avoid pregnancy and the oral contraceptive pill during this time. In 90% of patients the molar trophoblast regresses without requiring chemotherapy. Choriocarcinoma is fortunately a very chemosensitive tumour.

Finally, no matter what the type of pregnancy it is still a loss to the couple and this can easily be overlooked, particularly when major life-threatening complications have arisen. Women and their partners should be given ample time to express their feelings, air their anxieties and ask any questions. Couples should always be given the opportunity of talking through their feelings both at the time of their pregnancy loss but also at a later date when the woman may have recovered physically but will still need counselling to help her with her emotional trauma.

7.6.4   Recurrent miscarriage

Recurrent miscarriage is defined as three or more consecutive miscarriages. These may have occurred as pure chance. Often after careful investigation (see below), no cause will be found.

Investigation of recurrent miscarriage

Hysteroscopy
Chromosome analysis (both partners)
Thyroid function
Glucose tolerance test
Serum lupus anticoagulant
Serum cardiolipin
Lower genital tract infection screen

The chance of a successful pregnancy with no treatment is 60%. Immunotherapy with white cell transfusions has been attempted but is of debatable value as are attempts to improve progesterone levels either directly by giving progesterone or indirectly by stimulating the corpus luteum with HCG. Suppressing high levels of LH prior to ovulation may help to lower the high miscarriage rate in women with polycystic ovarian syndrome. Stress management is a promising treatment but as yet there are no comparative trials.

7.7   Conclusions

Infertility and pregnancy loss are both very common conditions and will therefore be seen frequently in the gynaecological outpatient department and on the gynaecology ward. Both conditions cause couples a great deal of distress and should be sensitively handled. Although treatments for infertility can help many couples, it is most important that couples should be given a realistic expectation of what any treatment can and cannot achieve and counselling should be used to help those who fail to achieve parenthood. As our understanding of reproduction grows so more couples can be helped to achieve a pregnancy and more women can continue with their pregnancy to term.

Learning points

Infertility affects 1 in 5 couples
Failing to conceive and/or pregnancy loss is associated with major psychological and social stress
Both partners should be seen, counselled and investigated as one unit
The initial history, examination and investigation should be directed at:
Confirming ovulation
Confirming adequate and normal sperm production
Confirming tubal patency
Confirming normal intercourse
Endometriosis, even minor degrees may cause subfertility
In a proportion of couples, no cause will be found
Ultrasound examination allows follicular development and ovulation to be observed (try to see this test done)

Further reading

Harrison, R. (1990) Stress in infertility, in *Recent Advances in Obstetrics and Gynæcology*, no 16 (ed. J. Bonnar) Churchill Livingstone, Edinburgh, London, Melbourne and New York, pp. 199–217.
Neuberg, R. (1991) *Infertility*. Thorsons, Wellingborough.

8 Fertility control

Masoud Afnan and Melanie Mann

8.1 Historical overview

8.1.1 Early society

Despite a naturally high mortality rate, abortion, infanticide and human sacrifice, the concept of pregnancy prevention has existed since earliest times – from rites of medicine men to diets, magic potions etc. Other more physiological methods have included prolonged lactation, prepubertal coitus, celibacy, withdrawal and various substitutes for sexual intercourse.

A Chinese medical text written about 2700 BC contains a prescription for an abortifacient. A prescription for various vaginal pastes written on papyrus dates back to about 1850 BC. A formula for a medicated tampon which released lactic acid, a known spermicidal, is found as long ago as 1550 BC. The oldest and probably the most effective form of contraceptive practice at that time was probably coitus interruptus (see Genesis 38:9).

The first scholarly work on fertility control was by Soranus (98–138 AD). He distinguished between contraceptives and abortifacients, discussed indications and contraindications, and described a number of techniques including the use of goat bladders as condoms. The widespread use of contraceptive practice in Europe is attributed to the spread of Islam. Islamic religious law does not forbid birth control or abortion before the fourth month of pregnancy. Again, the technique of coitus interruptus or Azl was the most widespread form of contraception. Unfortunately, although European physicians possessed contraceptive knowledge, the influence of Christianity prevented its practice. St Thomas Aquinas (1225–1274) condemned it as a vice against nature, and this has been re-affirmed through the centuries by a succession of popes, classically in the Humanae Vitae of 1968.

8.1.2   The development of family planning

Contraceptive practice is now available to a large majority of people in the developed world. Coordinated family planning programmes developed for a number of reasons. One was to enable individuals to space children for better personal health. Initially contraception was used to avoid responsibilities and costs of child-rearing. Only later did it become apparent that excessive child-bearing was injurious to health. Another was the concern for the consequences of high rates of population growth. Thomas Malthus, an English clergyman, wrote an essay in 1798 entitled 'The Principle of Population as it Affects the Future Improvement of Society'. In essence he theorized that unchecked population growth outstripped the means of its subsistence. This idea was overshadowed until the death rate diminished this century, and it once again appeared on the political agenda. The increase in total population as well as the decreasing age at menarche, combined with increasing age at marriage has resulted in an increasing population requiring contraception.

8.1.3   Technological innovations

Fallopius in 1564 described a linen sheath designed to cover the glans penis as a protection against syphilis. The vulcanization of rubber in 1844 eventually made possible the manufacture of condoms on a large scale.

The first cervical cap was made in 1838 by Wilde, a German gynaecologist. The first diaphragm to be made was by Hasse under the pseudonym of Mensinga in 1880. The first commercially available spermicidal cream was made in 1885, with significant modifications in the 1920s, and later in the 1950s.

Richter in 1909 described the first intrauterine contraceptive device (IUCD) and this was developed by Grafenburg in 1929.

The pill became available following isolation of synthetic oestrogens in 1954, and subsequent work published by Rock, Pincus and Garcia in 1956 and 1958.

The basic principles of contraception are as follows:

Principles of contraception

Abstinence
Suppressing ovulation
Preventing implantation
Altering cervical mucus
Altering endometrial receptivity
Occlusion of the fallopian tube, preventing transfer of the egg, sperm or embryo
Preventing the delivery of sperm   Barrier methods
                                     Vasectomy
                                     Spermicides

8.2   Physiological variations in fertility

Fertility is reduced in the first few months or years following menarche and also in the first few weeks after delivery. Continuing lactation causes relative infertility. There is also a decline in fertility prior to the menopause.

These periods are important, first, as an elegant example of nature reducing the risks of childbearing at times which carry greater than usual risk to the mother and child. Second, they cause statistical difficulties when assessing demographic data. Third, women may be misled into a false sense of security. Lastly, the choice of contraceptive may differ against the background of natural fertility.

Menses usually occur as early as 5–6 weeks after delivery. Ovulation occurs prior to the first period in 10–80% of postpartum women. It is much more likely if the woman is not breast feeding, though the protective effect of breast feeding decreases with time.

Fertility during lactation declines due to anovulation as a result of the high levels of prolactin, which in turn increase turnover of dopamine (prolactin inhibiting factor) which is also an inhibitory neurotransmitter for GnRH release. Lactation therefore delays the onset of menstruation, and even after affords some protection. At 12 months after the first postpartum period, cumulative pregnancy rates are 60% in breast feeding women as opposed to 80% in those not breast feeding.

Fertility declines as a function of age. In a group of previously fertile women 11% were unable to conceive by 35, 33% by age of 40 and 87% by 45. The choice of appropriate contraception in the older woman is made in the light of decreased fertility, risks of the contraception and the risks of pregnancy.

8.3   Effectiveness of contraception

There are two main methods for measuring the efficacy of a contraception:

Pearl index	Number of pregnancies per 100 woman-years. Does not account for key events such as discontinuation of use of contraceptive.
Life-table analysis	Used by insurance companies. Compares chance of a conception over a set period of time that the contraception is in use.

8.4   Hormonal contraception

The hormonal contraceptives are either oestrogen and progestogen or progestogens alone. The dosages of both have continued to decline over the past 30 years since they were first introduced. This has greatly reduced the health risks whilst maintaining efficacy.

8.4.1   Combined oral contraceptive (OC) pill

These are available as a fixed ratio of oestrogens to progestogens (monophasics), or varied in order to correspond roughly to the phase of the cycle (biphasics and triphasics). Ethinyloestradiol is by far the commonest oestrogen; the only other oestrogen in use is mestranol. The combined OC pill works by inhibiting ovulation. Additionally the progestogen component induces endometrial and cervical changes inhibiting implantation or sperm transport.

The dose of oestrogen should be the minimum required to keep the level of oestrogen just above the threshold for breakthrough bleeding. This can vary as achieved levels of oestrogens vary markedly among women taking the OC pill.

Side-effects of oestrogens are mainly those related to effects on the increased risk of thrombosis.

The type and dosage of progestogen will depend on the profile of the progestogen and the type of patient.

Types of progestogen in the combined pill	Dose
Norethisterone	0.5–1.5 mg
Levonorgestrel	150–250 µg
Ethynodiol diacetate	2 mg
Desogestrel	150 µg
Gestodene	75 µg
Norgestimate	250 µg

Progestogens such as levonorgestrel or norethisterone cause a slight depression in the cardioprotective high density lipoprotein 2 (HDL2) cholesterol levels. The newer progestogens such as desogestrel or gestodene do not have this effect. Progestogens generally, and levonorgestrel in particular, may cause insulin resistance and impairment of glucose tolerance.

Desogestrel, gestodene and norgestimate have little androgenic activity and should be used if the woman suffers from acne, hirsutism or weight gain.

Triphasics give better cycle control for a given low dose of progestogen, though they are more expensive. Additionally there is more room for errors in pill taking. They have also been noted to cause more premenstrual syndrome-type side-effects.

(a) Contraindications to combined oral contraceptive use

Circulatory disease	Arterial or venous thrombosis
	Ischaemic heart disease
	Hyperlipidaemia
	Severe arterial disease
	Coagulation abnormalities
	Conditions predisposing to thrombosis
	Transient ischaemic attacks
	Valvular heart disease
Liver disease	Acute liver disease
	Dubin–Johnson syndrome
	Cholestatic jaundice
	Porphyrias
	Liver adenomas
General	Undiagnosed genital tract bleeding
	Oestrogen-dependent neoplasms
	Trophoblastic disease
	Haemolytic uraemic syndrome

(b) Taking the OC pill

The OC pill should be started on the first day of the period. If the pill is started after day 4 of the cycle, extra precautions should be taken for the next 7 days. The most critical time for a pill to be missed is at either end of the pill-free week. If a pill is forgotten, it should be taken as soon as remembered, and the next one taken at the usual time. If 12 hours or more have elapsed, then extra precautions should be taken for the next 7 days. Should the 7 days run into the pill-free period, then the following pack should be continued without a break. If vomiting occurs within 3 h of taking the pill, then extra precautions should be taken. It is now advised that women, especially in younger age groups, use condoms with their sexual partners as well as using the oral contraceptive method. This has been shown to reduce the incidence of sexually transmitted diseases. It was pioneered in Holland and is known as the 'Double-Dutch method'.

(c)   Drug interactions and the OC pill

Liver enzyme inducers such as rifampicin, spironolactone, griseofulvin and all anticonvulsants interfere with all types of hormonal contraception. In these circumstances the OC pill should be taken as three packets in a row without a break, and only to take a 4 day break. There is more likelihood of breakthrough bleeding due to increased oestrogen metabolism, and so a higher oestrogen dose pill should be used. None of the high-dose oestrogen pills contain the newer progestogens. However, the newer pills can be combined to give the benefits of high dose oestrogens and the newer progestogens.

Broad spectrum antibiotics, particularly the penicillins and tetracycline, impair absorption of oestrogens, but not progestogens, by reducing the level of gut flora and thereby preventing enterohepatic recycling of the oestrogens. This will only be a problem in a small proportion of pill users, though as it is not possible to predict those in whom it is a problem, the advice is to use condoms for the duration of the course of antibiotics and for a further 7 days after the course has finished to ensure contraceptive safety. It is not a problem, however, for those on long-term antibiotic therapy as the gut flora renews itself. Enterohepatic circulation is not a mechanism in the maintenance of progesterone levels, as progestogen metabolites are inactive.

(d)   Cancer and the OC pill

There are definite links between the OC pill and the **prevention** of cancer, specifically ovarian (reduced by 40%) and endometrial (reduced by 50%) in long-term pill users. There is no relation to cervical cancer. The link with breast cancer, probably through the progestogen component is more confused. There may be a slight increase in women who start to take the pill when very young, but the latest studies looking at low-dose pills show no link with breast cancer.

8.4.2   Emergency (postcoital) contraception

This can be done in two ways – either by inhibiting ovulation, or by preventing implantation. The usual dosage is $100\,\mu g$ of ethinyloestradiol and $500\,\mu g$ of levonorgestrel (equal to two tablets of Ovran), repeated at exactly 12 h. The first tablets can be taken up to 72 h after the **first** episode of unprotected intercourse in that menstrual cycle. There is a failure rate of up to 4% at mid-cycle using this method. If there is an absolute contraindication to oestrogens, then $600\,\mu g$ of levonorgestrel (20 Microval tablets) can be given within 12 h of unprotected intercourse or an IUCD can be inserted. Nausea is the commonest side-effect with the high-dose combined pill, and if vomiting should occur within 3 h, then a further dose should be taken or the intrauterine contraceptive device (IUCD) inserted. A copper IUCD can be inserted up to 5 days after the calculated ovulation day, or in practice up to 5 days after unprotected sexual intercourse. In addition, an immediately effective form of contraceptive, such as a barrier method should be used till the next period. At this point future contraceptive practice is discussed and a follow-up visit should be arranged for 3–4 weeks later.

It is essential that all women using contraception, especially barrier methods are made aware of the availability of emergency contraception.

8.4.3 The progesterone only pill (POP)

The POP acts by making cervical mucus hostile and thick, thereby preventing sperm transport, and by inhibiting endometrial proliferation, thus preventing implantation. The POP also affects ovulation, but less reliably. In about 20% of women ovulation is inhibited with subsequent amenorrhoea. About 40% continue to ovulate normally, and 40% have some form of follicular development with subsequent irregular vaginal bleeding.

The POP is less effective than the combined OC pill, but by the age of 40 years, the failure rate is only 0.5 per 100 women years.

Femulen	Ethynodiol diacetate	500 µg
Micronor/Noriday	Norethisterone	350 µg
Microval/Norgeston	Levonorgestrel	30 µg
Neogest	Norgestrel	75 µg

The main indication for the POP is when the combined OC pill is contraindicated, such as in the older woman who smokes. It is also used during lactation as it does not inhibit lactation and it is very effective in this situation.

There are very few contraindications to the POP. The main problems are related to side-effects, such as irregular bleeding. There is also a very small increased risk of ectopic pregnancy. It should not be given to women with trophoblastic disease till HCG levels are undetectable.

The POP should be taken every day and within 3 h of the same time each day, and should be started on the first day of the cycle. The '7 day rule' also applies to the POP. Thus if a pill is missed by more than 3 h, condoms should be used for the following 7 days. It should be emphasized to the patient that this pill exerts its maximum effect about 4 h after taking the pill and so if intercourse usually takes place at night, the best time for pill taking is with the evening meal and not just before going to bed.

8.4.4   Injectable progestogens

The most widely used depot progestogen is medroxyprogesterone acetate (Depo-Provera) 150 mg, given every 12 weeks. Norethisterone enanthate (NEN ET), given every 8 weeks, is rarely used in the UK.

Both are given by a deep intramuscular injection and are formulated to release progestogen slowly into the circulation. The injection is given in the first 5 days of the cycle and the injection site should not be rubbed, as this may cause release of the hormone. If the drug is given in the puerperium, it should be given about 5 weeks after delivery. The method is highly effective, since in these dosages ovulation is inhibited, and the failure rate is the lowest among reversible methods (0–1 failures per 100 women years). Additionally compliance is not a problem – the method is reversible, and the median delay to the return of fertility is about 6 months. It is the method of choice for patients with sickle cell disease as it reduces the number of crisis attacks.

The main disadvantage is that if there are any side-effects, these will have to be tolerated for 3 months. The commonest side-effect is menstrual irregularity. Other problems may be weight gain or depression. The delay in return of fertility may also be a disadvantage to some women. The contraindications are similar to those of the POP though there is no increased risk of ectopic pregnancy. There is a slight reduction in serum HDL cholesterol, which in theory may increase the risk of cardiovascular disease. There is some evidence that women have become amenorrhoeic for several years with this method and thus will have low levels of oestradiol and, possibly, osteoporosis.

8.4.5   Progestogen-releasing silicone implants (Norplant)

Now available in the UK, these are inserted as six small subcutaneous rods under local anaesthetic. They are effective for 5 years, and can be removed if necessary. The failure rate is about 1 per 100 women years. The main disadvantage is that of irregular bleeding. The progestogen dose is low and therefore other side-effects are minimal. The progesto-gen-releasing vaginal ring (Femring) releases a low dose of progestogen over a 90-day period. The method is easily and quickly reversible, just by removing the ring. The dose of progestogen released is low, and the failure rate is 3 per 100 women years. Specific side-effects are vaginal irritation and discharge. Occasionally expulsion of the ring occurs, but it is simply replaced by the woman.

Recently a levonorgestrel-releasing IUCD (Mirena) has become available in the UK. It is a very reliable, safe method of contraception. The failure is 0.14 per 100 women years. There is a low incidence of ectopic pregnancy and fertility returns as soon as the IUCD is removed. The IUCD releases 20 μg of levonorgestrel daily once it has reached steady-state levels, and the main side-effect is irregular bleeding for the first few months. Menstruation then becomes much lighter and less painful and it has been advocated as a good method for the patients with menorrhagia.

8.5   Non-hormonal contraception

8.5.1   Intrauterine contraceptive devices (IUCD)

IUCDs have declined in usage since a peak of 6% of the fertile popu-lation in 1979. In many ways they are the ideal contraceptive in that they offer a high degree of reliability and effectiveness, their effect is quickly reversible, and there is no problem with patient compliance. However, fears over pelvic infection, irregular bleeding and lower abdominal pain have all served to limit its use. Additionally, a degree of skill is required to insert the IUCD.

Initially, the IUCD was an inert piece of plastic, such as the Lippes Loop (Figure 8.1a and b) or the Dalkon shield. The latter fell into disrepute as it was associated with serious infections, thought to be due to ascending infection via the multifilament thread. The second generation IUCDs were copper-bearing devices (Figure 8.2). As a result of the contraceptive effect of copper, the IUCDs were smaller, and therefore side-effects such as pain or bleeding were minimized. Ex-amples of the latter type of IUCD are Gravigard, Multiload and Ortho-Gynae T. They are all designed to fit the contour of the uterus and prevent expulsion, while minimizing the risk of perforation.

The IUCD is inserted either towards the end of a period when the

Figure 8.1 (a) Typical linear device – Lippes loop (left); typical closed device – ring (right). (b) Uterus with Lippes loop in place.

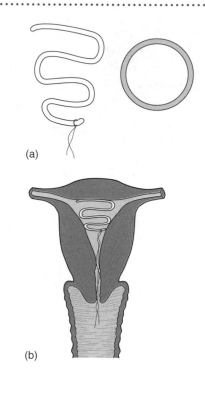

(a)

(b)

Figure 8.2 Copper-bearing IUDs: Tcu-200 (left) and Copper 7 (right).

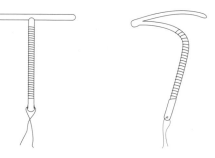

uterus and cervix are most relaxed and the patient is very unlikely to be pregnant (Figure 8.3). The woman should be examined after 6 weeks to ensure that the threads indicate the IUCD to be in the correct position. If the threads cannot be seen, then the IUCD has either been expelled or has been drawn up into the uterine cavity or even perforated through the uterus into the peritoneal cavity. The diagnosis is usually made by ultrasound scan or X-ray if necessary. Before performing any uterine procedure, however, pregnancy should always be excluded. The IUCD should then be checked every 6–12 months until time for replacement or removal.

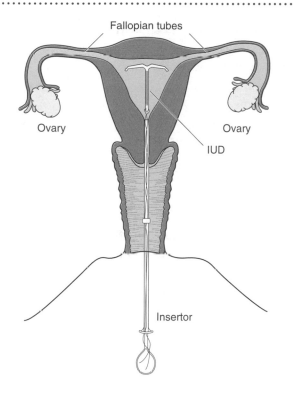

Figure 8.3 Inserting an IUD. Reproduced from *The IUD* published by The Family Planning Information Service.

Contraindications to IUCD insertion

Pregnant
Active pelvic infection or sexually transmitted disease within
    the preceding year
Undiagnosed abnormal bleeding from the genital tract
Uterine distortion due to congenital malformations or fibroids
Copper allergy or Wilson's disease

Relative contraindications
Nulliparity
Small uterus
Risk of bacteraemia unacceptable – Valvular heart disease
                                    Immunosuppression
                                    HIV seropositivity

Menorrhagia may become exacerbated within the first few months and care should be taken, especially if the patient is anaemic.

The main problems with the IUCD are heavy bleeding, pain, pelvic infection and ectopic pregnancy. Irregular or heavy bleeding is a common problem in the first few months, but usually settles down. However, about 10% of women will have the IUCD removed because of unacceptable bleeding patterns. Pain is often experienced at the time of insertion of the IUCD. If it persists, it may be that the stem of the IUCD is too long for the uterine cavity. If pain develops as a new symptom, the possibility of pelvic infection or ectopic pregnancy should be taken seriously. IUCDs are better at preventing intrauterine pregnancies than ectopic pregnancies. If a pregnancy occurs, the risk of an ectopic in a patient with an IUCD is 1 : 30 as opposed to 1 : 250 for non-users. There is, however, no convincing evidence that the IUCD causes ectopic pregnancy. The IUCD doubles the risk of pelvic infection, especially in the younger age group. Usually pelvic infection related to IUCD use presents within the first 4 months of insertion of the IUCD. Most cases respond to broad spectrum antibiotics. The device should be removed if the patient has failed to respond to treatment within 48 hours or if the infection is severe. If a woman has one steady sexual partner, the risk of pelvic inflammatory disease with the IUCD is very low.

IUCDs, especially the newer ones do not need replacing more often than every 5 years. The Multiload Cu375 is effective for at least 8 years. A device fitted in a patient age 40 or above does not need to be refitted.

The IUCD can be used for emergency contraception. It may be inserted up to 5 days beyond the calculated day of ovulation, and prevents implantation. It is particularly useful for those patients in whom the IUCD would be a suitable choice of contraception in any event.

8.5.2   Barrier methods

The main advantage of barrier methods is the absence of serious health hazards. If used correctly they provide effective contraception. They also have other benefits, such as minimizing the risk of contracting a sexually transmitted disease. They act either by preventing live sperm entering the cervical canal, or by killing the sperm.

(a)   Diaphragms and caps

Modern caps and diaphragms were developed in the nineteenth century, but their popularity has declined since the introduction of the pill in the 1950s. Recently there has been a return in popularity of barrier methods, due to concerns over the risks involved with the other forms of contraception as well as the non-contraceptive benefits, particularly protection from STDs and HIV. Diaphragms and caps are used along with spermicides for maximum efficacy. They have an advantage over the condom in that they are under the control of the woman and they do not have to be inserted at the height of sexual activity. Additionally spermicide can be added later as and when needed. However, they do not afford the same degree of protection against sexually transmitted disease. Recently, the female condom (Femidom) has been introduced (Figure 8.4). It is a disposable loose plastic sheath which covers the whole vagina and external genitalia (Figure 8.5a,b).

The diaphragm fits across the top of the vagina behind the pubic symphysis. It comes in a variety of sizes from 55 to 100 mm in 5 mm intervals (Figure 8.6). Care has to be taken in choosing the correct size (Figure 8.7) as well as teaching the patient how to fit the diaphragm (Figure 8.8a–c). If the diaphragm is too large (Figure 8.9) it will not be correctly positioned, and it may lead to excessive pressure on the vagina and possibly the bladder neck. If the diaphragm is too small (Figure 8.10), it will not maintain the correct position during intercourse. Prior to fitting the diaphragm, spermicidal cream should be applied.

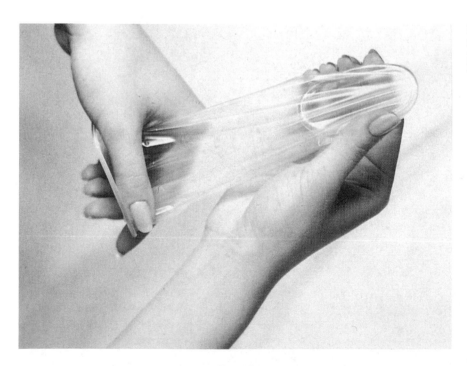

Figure 8.4 The female condom (Femidom). Reproduced with permission from Chartex International plc.

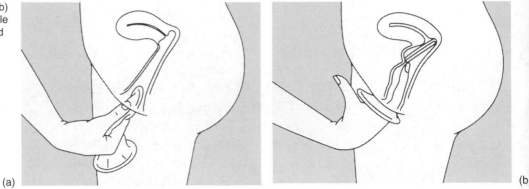

Figure 8.5 (a) and (b) Insertion of the female condom. Reproduced from *Handbook of Family Planning* by Nancy Louden.

(a)

(b

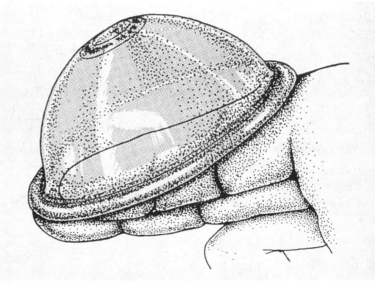

Figure 8.6 Diaphragm before insertion. Reproduced from *Handbook of Family Planning* by Nancy Louden.

Figure 8.7 Estimating the correct size of diaphragm to be fitted. Reproduced from *Handbook of Family Planning* by Nancy Louden.

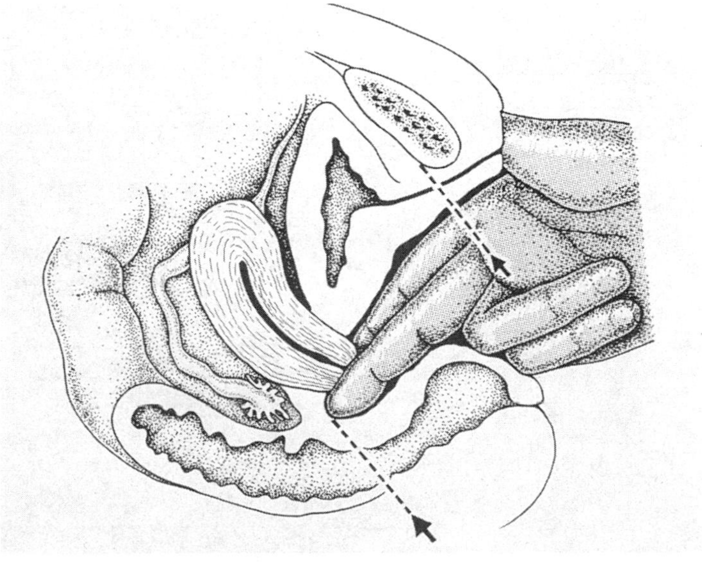

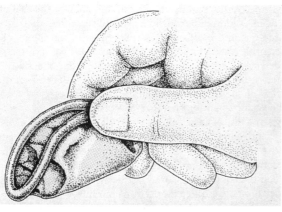

(a)

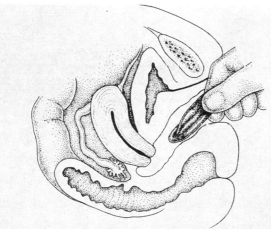

(b)

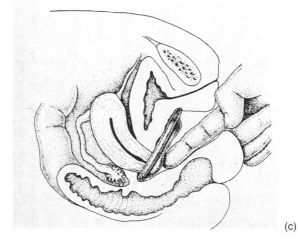

(c)

Figure 8.8 (a) and (b) Inserting a diaphragm. (c) Diaphragm correctly positioned. Reproduced from *Handbook of Family Planning* by Nancy Louden.

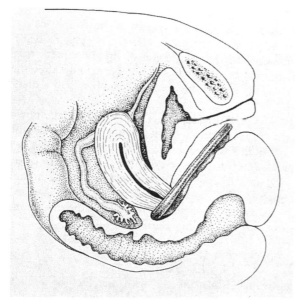

Figure 8.9 Diaphragm too large – incorrectly positioned in front of the pubic symphysis. Reproduced from *Handbook of Family Planning* by Nancy Louden.

Figure 8.10 Diaphragm too small – incorrectly positioned in front of the cervix. Reproduced from *Handbook of Family Planning* by Nancy Louden.

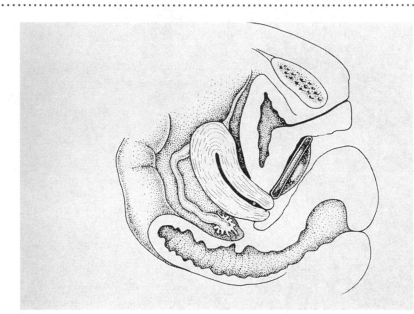

Figure 8.11 Cervical cap in position. Reproduced from *Diaphragms, Caps and the Sponge* published by The Family Planning Information Service.

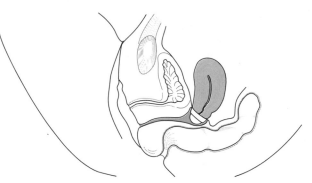

First year failure rates have been estimated at 2–3 per 100 users in one study, though other studies have shown a higher failure rate.

Cervical caps are less favoured by women. They are held in place over the cervix by suction (Figure 8.11). They are more often dislodged during intercourse, and so have a higher failure rate. They are useful if the woman has a problem with the diaphragm being retained and in whom the cervix is easily felt. As with diaphragms, spermicides are used with the cervical caps. Having had the cap or diaphragm fitted, the woman is asked to return after 6 weeks to ensure there are no problems. She should be asked to insert the diaphragm before visiting the clinic so that the examiner can tell if it has been correctly fitted. She should be reviewed every 6–12 months. If there are any changes in weight, or after childbirth or pelvic surgery, the cap or diaphragm may have to be refitted because of possible changes in pelvic anatomy.

Contraindications to diaphragm use

Congenital malformations affecting the vagina and cervix
Uterovaginal prolapse
Poor pelvic muscle tone
Allergy to rubber
Past history of toxic shock syndrome
Unable to fit the device herself

(b)   Spermicides

These are not very effective if used on their own. One year pregnancy rates of 17–22% have been reported in a large study. Most of the spermicides are nonoxynol based. They are available either as creams, gels, pessaries, foams, soluble film or incorporated in a contraceptive sponge. Occasionally spermicides may cause local irritation, which is usually due to the non-pharmacological agents in the preparation. An unperfumed gel (Gynol ll) may be used in these instances. The sponge is a polyurethane foam impregnated with 1 g of nonoxynol-9. It is moistened and inserted in the vagina to cover the cervix. There is a loop attached to facilitate removal. It is effective for 24 h. Pregnancy rates vary from 9 to 27 per 100 women.

(c)   The condom

This barrier method, also called the sheath, has been popular for two centuries. The availability and ease of use of condoms have made them a popular choice, both for short- and long-term use, until pregnancy is desired or until some other form of contraceptive is chosen. Most but not all condoms are lubricated with spermicides. Successful use depends upon the motivation of the couple. The condom should be applied over the erect penis before any genital contact is made. Before and during application the nipple at the head of the condom should be pinched to exclude air. This will help to minimize the risk of rupture – estimated as between 1 and 11%. After intercourse, the penis should be withdrawn before it gets too soft. The base of the condom should be held during withdrawal taking care not to spill any semen. It is estimated that about 13% of couples rely on the condom for contraception. Theoretical efficacy rates have been estimated to range from 0.5 to 2 pregnancies per 100 couple years. Its use is also to be encouraged in the prevention of sexually transmitted diseases.

8.6   Natural family planning

This depends on recognition by the couple of the fertile and infertile phases of the menstrual cycle, as well as an understanding of the female anatomy.

Coitus interruptus is probably the oldest method of fertility control, and it is still used widely. Reliable figures, including failure rates are difficult to obtain. Suffice it to say that it requires considerable restraint on the part of the male partner, and that notwithstanding, the semen which is expressed **prior** to ejaculation is also rich in sperm.

The so called 'rhythm method' can be used in women with a very regular cycle. It allows at least 5 days for sperm survival, and 48 h for ovum survival. The last six cycles are recorded and the longest and shortest defined. By subtracting 20 from the length of the shortest cycle, the first day of the fertile phase is identified. Similarly by subtracting 11 days from the longest cycle the last day is defined. So, if a woman notes that over 6 months her cycle varies from 26 to 30 days, her 'fertile phase' will be from day 6 to day 19 and the couple should avoid having intercourse during this phase.

The rhythm method can be combined with a number of other factors to improve accuracy and hence effectiveness. Due to rising progesterone levels, a rise in basal body temperature occurs at around the time of ovulation, and this can be used to define the end of the fertile phase, i.e. after the third consecutive day of a raised temperature. Under the influence of oestrogen, cervical mucus becomes more watery and is increased in volume at mid-cycle. In addition the cervix feels softer at mid-cycle. After ovulation, under the influence of progesterone, the cervix becomes firmer, with a reduced volume of thicker cervical mucus. These changes can be detected by the woman or her partner and used to detect the fertile phase. This method is known as the Billings method (after John Billings). However, more than 50% will cease using these methods within the first year of practice. Between 9 and 23% of women become pregnant in the course of 1 year of use. Recently 'home ovulation kits' have become available. They depend on the identification of the mid-cycle LH surge. The fertile period has already commenced by the time the LH surge takes place, and so these kits are not of benefit to those wishing to avoid pregnancy.

Methods of detecting the 'fertile phase'

Calculation from days of the cycle (rhythm method)
Increase in basal body temperature
Changes in cervical mucus
Changes in cervical consistency
Mid-cycle pain (in some women)

8.7   Sterilization

In the UK, the numbers of couples seeking permanent contraception has increased as concern is expressed about efficacy and safety of other methods of contraception. Careful counselling is required. It is important for the couple to realize that it is a permanent and irreversible form of contraception. Paradoxically, there is a failure rate associated with most forms of sterilization, and the couples should be aware of this also. The couple should realize that it is only fertility, and not sexuality, which will be affected. The two main forms of sterilization are vasectomy (male sterilization) and tubal occlusion (female sterilization).

Indications for sterilization vary from either patient choice of not wishing to continue contraceptive practices, to medical indications, where all other forms of fertility control are contraindicated. In the past, various inheritable conditions, such as Huntington's chorea, Tay–Sach's disease etc., have also provided the indication for sterilization, although as better contraceptive agents have become available, sterilization is rarely absolutely necessary. Certain patients with oestrogen-sensitive cancers, such as breast cancer and malignant melanoma have also been sterilized, though again, with better contraceptive agents available, this is less common. Sterilization of the mentally handicapped is an area fraught with controversy, but in certain situations can be another indication.

8.7.1   Vasectomy

This is the simpler of the two operations, often being done under local anaesthetic. Essentially the vas deferens is divided in the upper part of the scrotum as it passes upward from the epididymis. The main disadvantage is the delay in achieving sterility, which can take 3 months or longer. Sperm counts should be performed about 3 months after the procedure to ensure sterility, and some other form of contraception should be used until sterility is confirmed. Failure to render the patient sterile occurs in about 1% of cases. Complications of the procedure are rare, but include haematoma formation and, as a late complication, sperm granuloma formation. Vasectomy can sometimes be reversed, but the success of reversal depends on the length of time since the original procedure was performed, the development of antisperm antibodies and the surgical technique.

8.7.2   Female tubal occlusion

This is usually performed by laparoscopy under general anaesthetic as a day case and is immediately effective. Usually either clips (Filshie or Hulka–Clemens), or rings (Fallope) are applied across the tubes, or alternatively these can be ligated, excised or cauterized. If laparoscopy is contraindicated, then a mini-laparotomy may be performed. Alternatively, it may be performed during some other surgical procedure. The failure rate is increased if the procedure is done at the time of a Caesarean section, termination of pregnancy, or in the puerperium. During the first year, the failure rate is about 1 in 300. After this time, the failure rate drops to 1 in 1000. If the woman is having problems such as menorrhagia or dysmenorrhoea then hysterectomy might also be discussed as an appropriate way to deal with her problems and provide the family planning method of her choice.

However carefully counselled, a proportion of patients will return requesting a reversal of the procedure. The couple should be treated sympathetically, as no-one knows what the future holds. Nevertheless, there are a few situations in which a sterilization is performed and reversal sought more frequently. In these situations the couple should be very carefully counselled before agreeing to perform the procedure. The highest risks of this occurrence are in the young (under 30 years), the procedure being performed at the same time as a termination of pregnancy, and if there is marital discord.

8.8   Termination of pregnancy

There is probably no more controversial a subject in medicine than that of therapeutic termination of pregnancy. It is the most commonly performed procedure in the world. The ratio of live births to therapeutic abortion can be as high as 1 : 1 in some developed countries. In the UK, therapeutic termination of pregnancy became legal following the passage of the 1967 Abortion Act (Figure 8.12). Since then the numbers of therapeutic abortions have increased dramatically, with a co-existent decline in septic abortion with its attendant complications and fatalities (up to 1% of illegal abortions ended as a fatality). It is now the commonest female operation apart from episiotomy and it has been estimated that 1 in 3 women will terminate a pregnancy by the age of 30. Antibiotic prophylaxis has been proven to reduce the incidence of post-abortal pelvic inflammatory disease, caused mainly by *Chlamydia trachomatis*. Antibiotics should include anti-chlamydial drugs such as a tetracycline or a macrolide (e.g. erythromycin). Currently about 170 000 terminations are performed in the UK annually. The Abortion Act was revised in 1991 to lower the upper limit of gestation at which termination is permissible to 24 weeks, unless the fetus has been diagnosed as having a severe handicap in which case an upper limit is not stated.

Therapeutic termination of pregnancy can be performed surgically, or medically. In the first trimester (up to 12 weeks), vacuum aspiration can be performed from as early as the 6th week of gestation. In the second trimester, cervical dilatation and evacuation of the uterine contents can be performed, or the fetus and placental tissue expelled using either intra- or extra-amniotic prostaglandins. Other agents such as hypertonic saline or urea may be used as an intra-amniotic injection. After the abortion, a curettage of the uterus is usually required to remove adherent parts of the placenta.

Recently, an antiprogestin has been used to induce an abortion.

IN CONFIDENCE   **CERTIFICATE A**

ABORTION ACT 1967

Not to be destroyed within three years of the date of operation

**Certificate to be completed before an abortion is
performed under Section 1(1) of the Act**

I, ...
(Name and qualifications of practitioner in block capitals)

of ....

....
(Full address of practitioner)

Have/have not* seen/and examined* the pregnant woman to whom this certificate relates at

....

....
(full address of place at which patient was seen or examined)

on ....

and I ....
(Name and qualifications of practitioner in block capitals)

of ....

....
(Full address of practitioner)

Have/have not* seen/and examined* the pregnant woman to whom this certificate relates at

....

....
(Full address of place at which patient was seen or examined)

on ....

We hereby certify that we are of the opinion, formed in good faith, that in the case

of ....
(Full name of pregnant woman in block capitals)

of ....

....
(Usual place of residence of pregnant woman in block capitals)

(Ring appro-priate letter(s))	A	the continuance of the pregnancy would involve risk to the life of the pregnant woman greater than if the pregnancy were terminated;
	B	the termination is necessary to prevent grave permanent injury to the physical or mental health of the pregnant woman;
	C	the pregnancy has NOT exceeded its 24th week and that the continuance of the pregnancy would involve risk, greater than if the pregnancy were terminated, of injury to the physical or mental health of the pregnant woman;
	D	the pregnancy has NOT exceeded its 24th week and that the continuance of the pregnancy would involve risk, greater than if the pregnancy were terminated, of injury to the physical or mental health of any existing child(ren) of the family of the pregnant woman;
	E	there is a substantial risk that if the child were born it would suffer from such physical or mental abnormalities as to be seriously handicapped.

**This certificate of opinion is given before the commencement of the treatment for the termination
of pregnancy to which it refers and relates to the circumstances of the pregnant woman's
individual case.**

Signed ....   **Date** ....

Signed ....   **Date** ....

* Delete as appropriate   Printed in the U.K. for H.M.S.O. 5/91 Dd. DH001306 C10000 38806 G3994   Form HSA1 (revised 1991)

Figure 8.12 Form HSA1 Abortion Act 1967 Certificate. Crown copyright is reproduced with the permission of the Controller of HMSO.

Known as RU-486 or Mifepristone, it acts by preventing or disrupting implantation. Efficacy is increased by combining it with prostaglandins. Success rates decline with advancing gestation. However, up to 95% of pregnancies of less than 8 weeks gestation are successfully terminated using prostaglandin E_2 and Mifepristone. The remaining 5% have to be terminated surgically.

Complications of terminations of pregnancy

Haemorrhage
Uterine perforation
Infection
Retained products of conception
Anaesthetic complications
Infertility due to tubal damage
Intrauterine adhesions (Asherman's syndrome)
Psychiatric after effects
Traumatic damage to the cervix
Rhesus isoimmunization

Death may also occur, with an incidence of about 1 in 100 000 abortions, usually due to haemorrhage, infection, pulmonary embolus or anaesthetic complications.

8.9 Future developments

Contraceptive vaccines against HCG, zona pellucida and sperm are all being developed. The anti-HCG vaccine is at the most advanced stage of evaluation. Considerable advances have taken place in the field of male contraception. The use of gonadotrophin releasing hormone agonists and antagonists, in combination with testosterone seems hopeful. Intravasal injections of occlusive agents, such as silicon are also being evaluated.

8.10 Conclusions

Control over fertility has always been important to humankind and more latterly to couples and individuals. It is a woman's right to determine when she will be pregnant and this, perhaps more than any other medical advance has provided women with a freedom that in the past they did not have.

Many methods of contraception are available and developments continue to refine the techniques making them both more effective and safer.

It is always important to counsel carefully any woman or couple seeking contraceptive advice. Most failures that occur are user failures and not method failures therefore adequate explanation and appropriate choices will have a major bearing on compliance and therefore success or failure.

Learning points

There are normal physiological variations in fertility

The couple's needs and wishes need adequate discussion before choosing an appropriate method

The basic principles are prevention of ovulation, preventing implantation and preventing contact between sperm and egg

Sterilization is a permanent method but does have a failure rate

Barrier methods, particularly the condom, provide added protection against sexually transmitted diseases

Natural methods of family planning require basic understanding and motivation on the part of the couple

You should know the side-effects, indications and contraindications of hormonal contraception and IUCDs

---

Further reading

Guillebaud, J. *Contraception: Your Questions Answered*, 2nd edn. Churchill Livingstone, Edinburgh.

Guillebaud, J. *Contraception: Hormonal and Barrier Methods*. Martin Dunitz, London.

Loudon, N., Glasier, A. and Gebbie, A. (eds.) *Handbook of Family Planning and Reproductive Healthcare*. Churchill Livingstone, Edinburgh.

9 The abnormal cervical smear

David Luesley

9.1 Introduction

Cervical smear tests are performed with the intention of detecting the asymptomatic precursors of cervical cancer. The current consensus is that cancer of the cervix can be prevented by early detection and treatment of preinvasive disease.

In the United Kingdom approximately 4000 new cases of cervical cancer are diagnosed each year. Populations that have been subjected to rigorous screening programmes have documented falling incidence of the disease. Although this may be attributable to effects other than screening most authorities believe that cervical cytology screening has made an impact and is a prevention strategy worthy of continuing.

The test was initially devised by Papanicolau and was designed to detect the preinvasive phase of squamous cancer. This is cervical intraepithelial neoplasia (CIN). A proportion of cervix cancers are not squamous such as adenocarcinoma of the cervix. Some of these cancers also have a precursor (cervical glandular intraepithelial neoplasia or CGIN). These conditions can also be detected on a smear test although less reliably.

Population screening is organized in some countries and in the UK a three to five-yearly smear is recommended for all women. Screening usually begins in the early twenties and continues until the age of 65 years.

As with any screening strategy, a knowledge of the aetiology and natural history of the condition is essential.

9.2   Aetiology

There is strong circumstantial evidence to implicate a sexually transmitted factor in the genesis of CIN. This is based on the observed increased relative risk in those exposed to multiple sexual partners from an early age. Human papillomaviruses (HPV), particularly types 16 and 18 are closely related with malignant transformation and they may be one of a group of potential oncogenic influences. Herpes viruses have also been associated with malignant disease of the cervix as have various components of seminal fluid. Smoking is also thought to be implicated perhaps via an immunosuppressive effect. Much has been written in the past regarding early coitus and multiple partners and as such the disease has become associated with promiscuity. This approach is counterproductive for although sexual habit undoubtedly influences cervical pathology it is also true that the majority of women who have abnormal smears do not have sexual habits at variance from the normal population and the continued association of the two only results in the perpetuation of sexual stigma. Some recent evidence has suggested the concept of a high risk male partner. This may explain the increased frequency of cervical cancer seen in the subsequent consorts of men who have had a consort with cervical cancer. It also supports the concept of a sexually transmitted factor such as HPV.

Factors possibly involved in cervical carcinogenesis

Human papillomaviruses (types 16, 18, 31 and 33)
Herpes viruses
Seminal fluid
Cigarette smoking
Immunosuppression

9.3   The cervical transformation zone

The malignant changes, whatever their cause, occur in the area of metaplasia lying between glandular endocervical epithelium and squa-

Figure 9.1 Diagrammatic representation of the cervix showing migration of the squamocolumnar junction (SCJ).

Squamocolumnar junction sited in the endocervical canal

Expansion and eversion of the cervix under the influence of oestrogen results in an ectocervical position of the SCJ

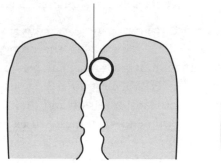
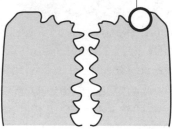

mous ectocervical epithelium. This area is normally sited at the external cervical os and is amenable to surface scraping (the smear).

The shape and consistency of the cervix changes throughout life under the influence of endogenous and exogenous oestrogen. This effect may cause the position of the junction between squamous and glandular epithelium to move (expansion and eversion under the influence of oestrogen and contraction and inversion on its withdrawal (Figure 9.1). The low pH in the vagina is the stimulus to squamous metaplasia, thus as the cervix expands and everts under the influence of oestrogen, glandular epithelium is exposed to a lower pH and undergoes metaplastic change. If this area becomes exposed to oncogenic influences, the metaplastic process becomes dysplastic. Dysplasia is another term for CIN. It therefore becomes apparent that the smear test should aim to sample the area of metaplastic transformation and this is the junction between the two epithelia.

CIN can affect the stratified epithelium of the transformation zone in varying degrees (Figure 9.2). If all of the layers show evidence of malignant transformation (reduced nuclear cytoplasmic ratio, mitotic figures etc.) then this is called CIN 3 or high-grade disease (c). If only the most basal layers show any degree of change then this is CIN 1 (a). CIN 2 is an intermediate level (b). It will be appreciated that to make this type of diagnosis a full thickness biopsy will be required thus although a smear might indicate an abnormality it is not possible to make the diagnosis on a smear alone. High-grade disease (CIN 3 and CIN 2) probably has the potential to develop into cancer if left untreated. The risk of progression can only be estimated but between 30 and 40% of such cases may become invasive over 20 years. The rate of

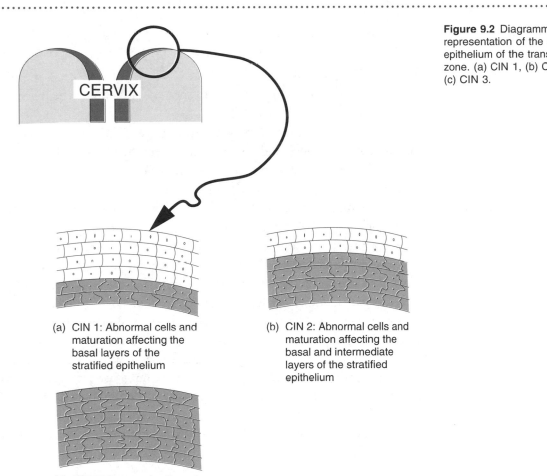

(a) CIN 1: Abnormal cells and maturation affecting the basal layers of the stratified epithelium

(b) CIN 2: Abnormal cells and maturation affecting the basal and intermediate layers of the stratified epithelium

(c) CIN 3: Abnormal cells and maturation affecting all layers of the stratified epithelium

Figure 9.2 Diagrammatic representation of the stratified epithelium of the transformation zone. (a) CIN 1, (b) CIN 2 and (c) CIN 3.

progression applies to populations and not individuals where wide variations in progression rate might be observed. It is because of this that 3–5-yearly tests are advocated. This strategy will confer about a 90% protection level on a population assuming that the natural history of the disease does not change.

Whether low grade (CIN 1) will become high grade eventually and then progress to cancer is much more dubious. Nevertheless, as it is not as yet possible to determine the risk accurately in this group the tendency has been either to maintain careful cytological surveillance (smears 6 monthly) or to offer treatment.

9.4   Taking a smear

The ideal time to take a smear is mid-cycle and at least 3 months after any pregnancy. Smears should not be taken at the time of menstruation.

A careful explanation of how the test is taken helps to relax the woman although most find the experience embarrassing. The woman should relax in the dorsal position, be minimally exposed and ensured of privacy. The cervix must be fully visualized, this is usually best achieved using a warmed lightly lubricated bivalve speculum. A small amount of a water-based lubricant does not contaminate the smear so as to render interpretation difficult. Any heavy discharge or blood (say from a traumatized ectropion) should be gently swabbed away. The actual appearance of the cervix may be important and if at this stage there is a suspicion of malignancy then this in itself should prompt further investigation regardless of the smear result. If the junction between the glandular and squamous epithelium can be seen then this should be the target of a 360 degree firm surface sweep with a suitable spatula (Figure 9.3). Several types of spatula have been designed. The design objective has been to enhance the harvest of endocervical cells as it has been suggested that this indicates that the transformation zone has been adequately sampled. In situations where the junction is felt to be sited within the endocervical canal there may be some justification

Figure 9.3 Diagrammatic representation of obtaining a cervical smear using a spatula.

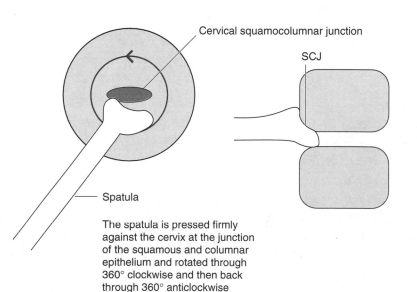

Cervical squamocolumnar junction

SCJ

Spatula

The spatula is pressed firmly against the cervix at the junction of the squamous and columnar epithelium and rotated through 360° clockwise and then back through 360° anticlockwise in using a spatula that can access the canal. It is far more important, however, to fully visualize the cervix, collect a good cellular sample, evenly spread it over a labelled glass slide and fix it rapidly with ethanol. Sometimes the act of taking a smear may cause some bleeding, particularly if an ectropion is present. The woman should be reassured that this is quite normal and that the bleeding will not be heavy and will rapidly subside. *Learning to take a cervical smear should be regarded as an essential clinical skill.*

9.5   Explaining the test to women

Most women having smear tests are aware that the procedure is related to cancer of the cervix, indeed many refer to it as 'A Cancer Smear'. This in itself is the source of most of the anxieties that women suffer. It is important to explain prior to taking the test why it is being done. In this context the idea of prevention rather than diagnosis of cancer cannot be underestimated. Terms such as precancer, dysplasia and CIN only confuse the majority and heighten their underlying fears. It is therefore important to use understandable terminology. 'The test is to take a sample of cells from the skin of the cervix (neck of the womb) to see if in the future there might be a risk of developing cancer of the cervix.' This is a more understandable explanation than . . . 'The smear test will detect precancer'.

Together with an explanation of why the test is being done, women require additional information at the time of the test. What happens next, i.e. what will happen if the test is abnormal and what will happen if the test is normal. Similarly some idea of when the result will be available and how they will be informed should be given at the time of the test. There should be ample opportunity to ask questions and counselling regarding these aspects of the test should be performed before or afterwards, not while the woman is in the most vulnerable (and most anxious) position of having the test performed.

9.6   The results

Cervical smears are read manually. Results should be in a format agreed by the British Society for Cervical Cytology.

They are reported as follows:

Result	Interpretation
Negative	A normal smear (Figure 9.4a)
Borderline nuclear abnormalities	Minor increase in the nuclear cytoplasmic ratio but insufficient to be classed as dyskaryotic
Mild dyskaryosis	Abnormal, suggestive of, but not diagnostic of mild dysplasia (CIN 1) (Figure 9.4b)
Moderate dyskaryosis	Abnormal, suggestive of, but not diagnostic of moderate dysplasia (CIN 2)
Severe dyskaryosis	Abnormal, suggestive of, but not diagnostic of severe dysplasia (CIN 3) (Figures 9.4c and 9.5)
? Invasion	Abnormal, smear shows evidence of a malignant diathesis (fibre cells or necrotic debris)
Koilocytosis	Cells suggestive of infection with human papillomavirus (Figure 9.4d)
Inadequate	Cellular content either insufficient or smear unsatisfactorily prepared to allow for a cytological opinion.

Occasionally a report will be seen which indicates inflammatory changes. This result is now discouraged unless a specific pathogen causing such an inflammatory response can be identified. This may be monilia or trichomonas. Occasionally herpes or actinomycetes may be reported. Although cytology should not be regarded as the diagnostic basis for these infections, the inflammatory changes they cause may mask underlying dyskaryosis and the infection should be treated and or the smear repeated once the inflammation has subsided.

Whatever the result the woman should be informed of it. In the case of an abnormal result this is best done at a consultation in order that a full explanation can be given with particular regard to allaying anxiety

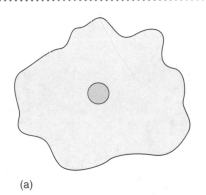

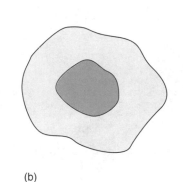

(a) (b)

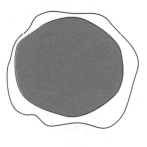

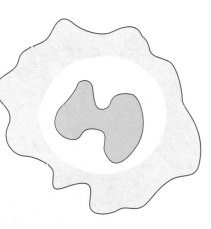

(c) (d)

Figure 9.4 Diagrammatic representation of smear test results. (a) A normal squamous cell; (b) mildly dyskaryotic cell; (c) severely dyskaryotic cell, suggestive of CIN 3; (d) koilocyte with a bilobed nucleus and perinuclear halo. This type of cell is suggestive of a human papillomavirus infection.

and explaining what will happen next. It is important to remember that the smear result is not diagnostic and some abnormalities, particularly minor changes may revert to normal without intervention.

9.7   Who should be referred?

All patients who have a severely dyskaryotic smear or worse should be referred for further investigation as there is a high probability (approximately 80%) of their having high grade dysplasia (CIN 2 and CIN 3). Similarly any patient with a clinically suspicious cervix (heavy contact

Figure 9.5 A cervical smear showing severe dyskaryosis. Note the sharp difference between the normal squamous cell (large cell with small nucleus) and the severely dyskaryotic cells (small darkly staining cells with relatively large nuclei).

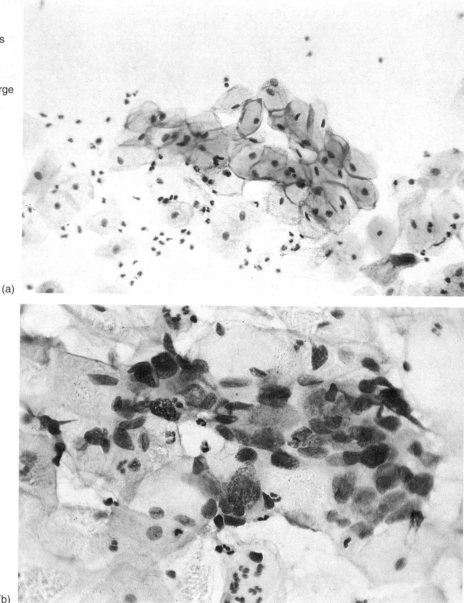

(a)

(b)

bleeding, irregular surface, necrotic debris, ulcer etc.) should also be referred. Most authorities would also recommend referral for moderate dyskaryosis although proportionally fewer will have high grade dysplasia. The situation is somewhat different for women with milder abnormalities on their smear (borderline nuclear abnormalities and mild dyskaryosis). Although about one third of these patients will harbour high-grade disease the remainder do not, and although the invasive potential of low-grade dysplasia or wart virus infection alone is unknown it is thought to be very low. There is therefore scope for allowing time for spontaneous resolution of the smear. A repeat smear in 6 months is appropriate management in these situations unless there are other reasons such as a suspicious cervix that might prompt referral. Approximately 40% of women who have a mildly dyskaryotic smear will have a normal smear 6 months later and an even greater proportion of those with only borderline changes. If the smear remains abnormal at a subsequent visit then referral should be considered. All patients who have had an abnormal smear should remain on heightened cytological surveillance and if not referred have at least three consecutive negative smears at annual intervals before being returned to the cytology screening programme. These patients will need considerable positive support to prevent the anxiety that often occurs with repeated minor cytological changes and close cytological surveillance.

Situations requiring further investigation

Moderately dyskaryotic smear or worse
Two consecutive minor abnormalities (6 months apart)
Suspicious looking cervix
Two consecutive inadequate smears (or severe inflammatory)
Anyone with an abnormal smear who has had previous
    treatment for CIN

9.8   Further investigation

Patients are referred in order to achieve an histological diagnosis and treatment, if indicated. This is achieved through colposcopy. This technique provides a magnified (×6–×20) and illuminated image of the cervix. The surface characteristics of the cervix, in particular the area of epithelium between the glandular endocervical and native squamous epithelium can be examined (Figure 9.6). This area, the transformation zone, contains metaplastic and dysplastic epithelium. If dysplasia is present some degree of characterization is possible utilizing such fea-

Figure 9.6 A normal cervix showing the junction between glandular (endocervical) and squamous epithelium.

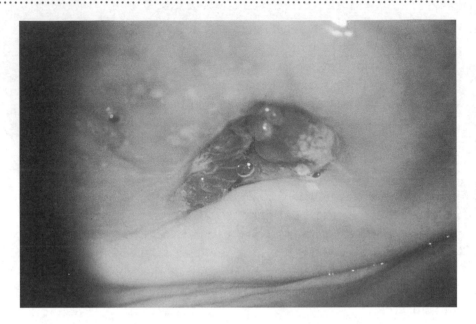

tures as the subepithelial vascular pattern (the more coarse and disordered the vessel pattern, the higher the degree of dysplasia). These capillaries appear as a mosaic or punctate pattern. The application of 5% acetic acid enhances these features and allows areas of metaplasia and dysplasia to stain white (acetowhite epithelium) (Figure 9.7). It also allows the size and surface contour of lesions to be characterized (Figure 9.8). The colposcopist will select the area with the most abnormal changes and take a directed biopsy. Alternatively, if any abnormality is seen in the transformation zone it can be excised completely under local anaesthetic by a fine diathermy wire loop (Figure 9.9) or by using a laser beam. The specimen can then be histologically graded. If high grade CIN is confirmed by the directed biopsy a further procedure will be required to treat the whole transformation zone as the whole area is considered to be at risk of neoplastic change. If the initial biopsy was excisional, i.e. removed the whole transformation zone, then this will have been therapeutic as well as diagnostic.

If the upper limit of the transformation zone cannot be visualized as it may be sited well up inside the endocervical canal (particularly in postmenopausal women) it will be necessary to take a cone biopsy. A cone biopsy is a conical tissue biopsy that aims to remove all of the visualized lesion on the ectocervix and at least two thirds of the endocervical canal (Figure 9.10). They usually average 2–2.5 cm in length and are cut by either a scalpel blade (Knife Cone) or latterly with

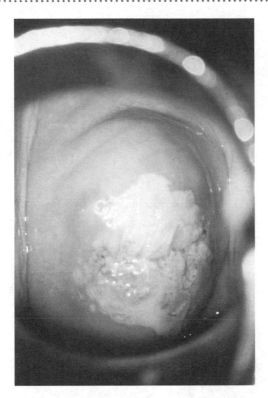

Figure 9.7 Cervix seen at colposcopy after the application of 5% acetic acid. Dysplastic epithelium (CIN) shows as the whitened area.

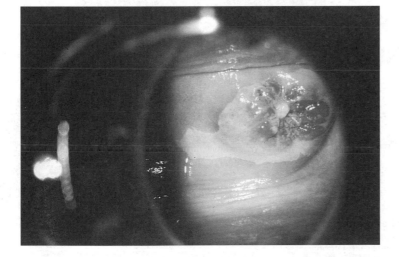

Figure 9.8 Colposcopy after the application of 5% acetic acid. In this case abnormal epithelium is seen extending onto the vagina. A biopsy of this area will be required to differentiate dysplasia from the 'congenital transformation zone'.

Figure 9.9 Diagram illustrating the technique of large loop excision of the cervical transformation zone and the type of excisional biopsy achieved. Cone biopsies are similar but usually remove much more of the endocervical canal. A knife or laser can be used to cut specimens like this.

TECHNIQUE

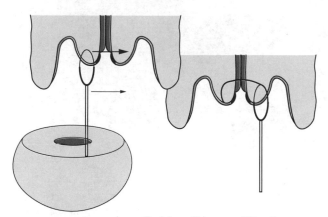

Loop Excision : Theory and Practice

Figure 9.10 Diagrammatic representation of a cone biopsy of the cervix.

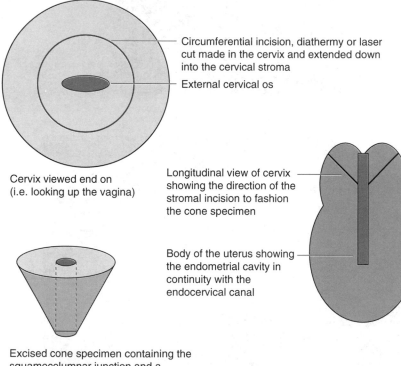

Circumferential incision, diathermy or laser cut made in the cervix and extended down into the cervical stroma

External cervical os

Cervix viewed end on (i.e. looking up the vagina)

Longitudinal view of cervix showing the direction of the stromal incision to fashion the cone specimen

Body of the uterus showing the endometrial cavity in continuity with the endocervical canal

Excised cone specimen containing the squamocolumnar junction and a segment of the endocervical canal

either a laser beam or large diathermy wire loop. This usually requires admission to hospital and a general anæsthetic although techniques are now being developed to allow this slightly larger biopsy to be removed under local anaesthetic in a clinic setting.

If the initial biopsy only confirmed low-grade dysplasia some might elect to observe the situation by further colposcopy, cytology and biopsies as indicated. Another school of thought, however, believes that once CIN, regardless of grade, has been confirmed then the whole transformation zone should be treated as the initial biopsy might be subject to a sampling error.

Treatment methods for CIN (cervical intraepithelial neoplasia)

Excisional	Destructive (Ablation)
Diathermy loop excision	Laser vaporization
Laser excision	Cold coagulation (destruction with a
Knife cone biopsy	hot probe (120–140°C)
	Diathermy
	Cryocautery (freezing)

Destructive methods require biopsy confirmation of CIN prior to treatment lest they are inadvertently applied in situations where invasive disease has already occurred. All of these local methods of treatment would be considered as inadequate to treat frank invasive disease of the cervix.

9.9 Counselling the woman

Colposcopy can be emotionally traumatic for many women given that they already fear they have a significant problem. Privacy, carefully worded and clear explanations are essential components of the consultation. Anxiety may hinder understanding and explanatory leaflets may be of use here. It may also be useful for patients to bring a friend or relative and if the woman wishes, involve them in the explanation. Visual aids such as a video camera attached to the colposcope are invaluable teaching aids and may help some but not all patients. Explaining what is happening, in simple terms is often useful although it is important to avoid medical jargon. (This confuses and is nearly always interpreted as something bad!) If treatment is contemplated, the woman should be told well in advance what the after effects will be.

These include a discharge, some light bleeding and occasionally some heavy bleeding. There is usually no pain after treatment and if adequate local analgesia is used treatment itself is painless. All treated women should refrain from intercourse and tampon-use for at least four weeks after treatment. (It is important to inform them of this prior to treatment vis-à-vis weddings, holidays, etc.). In the majority of situations the cervix heals in such a way as to restore normal cervical anatomy. Larger biopsies may result in excess scarring and occasionally stenosis of the cervical canal. This in turn may lead to menstrual dysfunction and dysmenorrhoea. There is also a possibility that cervical scarring may result in infertility or miscarriage and problems in subsequent labours. These potential problems remain theoretical at present as no data have confirmed a significant clinical effect. Nevertheless all possible outcomes should be discussed in the light of current knowledge.

9.10   Follow-up

The vast majority of cases of CIN (over 90%) are satisfactorily treated at the first attempt in that subsequent smears and colposcopic evaluations are normal. If the smear fails to return to normal, or there remains a colposcopic suspicion of CIN then a further treatment might be necessary. Women are seen 6 months after treatment, if the smear and colposcopy are normal they can usually be discharged to cytological surveillance. Current recommendations are that a smear should be performed annually for five years, if all remain negative three yearly routine smears should be performed. If any post-treatment smear is reported as abnormal then a repeat colposcopic assessment with biopsy if necessary, is indicated.

9.11   Microinvasion

Occasionally the biopsy will indicate that the malign epithelium has breached the basement membrane. This is no longer CIN. Colposcopy might suspect the presence of early invasion but by no means can it be regarded as accurate. Staging of early cervical cancer is totally dependent on histology. If the malign epithelium extends to 3 millimetres or

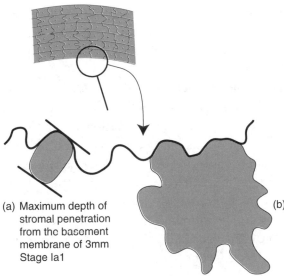

Figure 9.11 Diagrammatic representation of early cervical cancer (stage Ia).

(a) Maximum depth of stromal penetration from the basement membrane of 3mm Stage Ia1

(b) Penetration of greater than 3mm, but less than 5mm with a maximum lateral dimension of 7mm is Stage Ia2. Anything more than this is Stage Ib

less below the basement membrane then this is staged as Ia1 (Figure 9.11a). If the depth of penetration is greater, but still less than 5 mm and less than 7 mm across then this is staged as Ia2 (Figure 9.11b). Anything more than this is Stage Ib. The incidence of lymphatic involvement increases from about 2% with Stage Ia2 to in excess of 15% with Stage Ib. Because of this, different treatment approaches are required. Careful counselling approaches are required here. In Stage Ia1 and Ia2, although the risk of disseminated disease is small it is still possible, therefore, unless future childbearing is a major issue further treatment might well be justifiable particularly if there are doubts as to whether the whole area of abnormality has been removed. In older women hysterectomy is often the treatment of choice with removal of the pelvic lymph nodes in selected cases of Stage Ia2 (i.e. if neoplastic cells are seen lying within capillary-like or lymphatic channels).

9.12   Hysterectomy in the treatment of CIN

In addition to those cases of early invasion there may be situations where hysterectomy is the best option to treat confirmed CIN. These would normally be associated gynæcological problems such as uterine fibroids, menorrhagia etc. Again this should be part of the initial counselling process and underlines the need for accurate history taking even when an apparently asymptomatic patient presents with an abnormal smear. The woman's attention may well have been diverted away from what she perceives as something minor in relation to the relatively worrying abnormal smear.

If hysterectomy is contemplated there is still a need to assess the cervix and upper vagina colposcopically. A small proportion of women (about 4%) will have extension of the transformation zone onto the vaginal vault. Some of these extensions are quite benign and represent a very minor congenital abnormality (the congenital transformation zone). There may, however, be dysplastic change in the upper vagina; if unrecognized potentially malign tissue might be left or buried in the vaginal vault at the time of hysterectomy. This can lead to a situation of VaIN (vaginal intraepithelial neoplasia). It is for this reason that women who have had a hysterectomy for known CIN are offered follow-up vaginal vault cytology. If the vault smear remains negative on two annual reviews and the histology of the hysterectomy specimen confirms complete excision they might safely discontinue cytological follow-up. Women who have had hysterectomies without any evidence of current or previous CIN do not require vaginal vault cytology. Any woman who is to have a hysterectomy for any benign condition should have a confirmed normal smear before doing so and if not normal should be colposcopically assessed prior to her operation.

9.13   Contraception and CIN

Most women who present with an abnormal smear will be in their childbearing years. One frequent anxiety is the possible effect of the contraceptive method on either the genesis or progression of CIN. Although theoretically, abstinence and barrier methods of contraception may reduce the risk of any sexually transmitted aetiological agent there are no data available that would suggest that a change in contraceptive method would minimize the risk of progression of disease. All women with abnormal smears should be counselled with regard to family planning and the major consideration should be which method of family planning the woman requires and/or wishes to continue using. The presence of an abnormal smear should not influence the choice of contraception. An unplanned pregnancy occurring as a result of fear of possible involvement of contraceptive method should be regarded as a major error of counselling.

It may be necessary to temporarily remove an intrauterine device at the time of treatment. This is mainly because of the risk of damaging the IUCD strings although the potential for causing or exacerbating pelvic infection as a result of causing a large healing crater cannot be discounted.

9.14 The pregnant woman

In some women, cytological abnormality is recognized in pregnancy. The same criteria for colposcopic referral operate in pregnancy as in the non-pregnant state. Colposcopy can be technically more difficult during pregnancy as one of the effects on the cervix is to accentuate the ectropion and increase the area of potential metaplastic change. Access to and good visualization of the cervix may also be more difficult. Biopsy should be avoided in pregnancy because of the increased vascularity of the cervix but may be necessary if an invasive process is suspected on either cytological or colposcopic grounds. If, after a thorough colposcopic assessment it is still not possible to exclude invasive disease a large and representative biopsy should be taken. A wedge of tissue, encompassing the most atypical areas of epithelium usually suffices to provide a suitable sample on which to base a diagnosis of invasion. Because of the possible risk of haemorrhage this is usually performed under a general anaesthetic. Wedge biopsy has not been associated with an increased risk of miscarriage. If intervention is not felt to be justified during pregnancy, either a further smear and or colposcopic assessment should be performed not before 3 months following delivery. This allows ample time for involution to occur. The presence of CIN cannot affect the fetus nor does pregnancy affect the course of CIN.

9.15 The postmenopausal woman

Women over the age of 65 years who have had regular and normal screening are at very little risk of developing cervical cancer. The screening programme therefore stops at this age. However, cancer of the cervix is still more prevalent in older women and screening is maintained well into the menopause.

Figure 9.12 Colposcopy in a postmenopausal woman. The cervix is atrophic, flush with the vaginal vault and the endocervical canal cannot be seen. Note the patchy almost haemorrhagic appearance of the squamous epithelium.

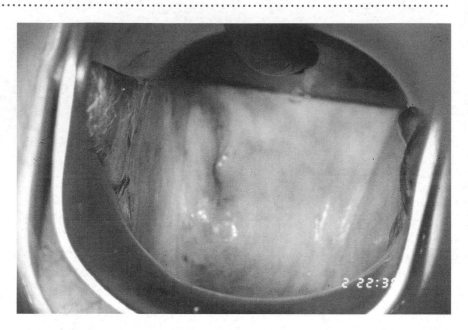

As a result of oestrogen loss profound changes affect the genital tract including the cervix. There is an overall shrinkage of the tissues with an inversion of the cervix and as a consequence the squamocolumnar junction may appear to recede up the endocervical canal (Figure 9.12). In addition to these changes the epithelium becomes thinner and more easily traumatized, there is less glycogenation of squamous cells and the pH of the vagina increases removing one level of vaginal defence. Smears from postmenopausal women therefore contain fewer cells, less mature squames and are less likely to adequately sample the squamocolumnar junction. Unsatisfactory samples and those reported as showing minor abnormalities appear to be reported more frequently in postmenopausal women. There would seem to be good theoretical grounds for providing oestrogen replacement in such women and then repeating the smear. Such a strategy has not been evaluated in clinical practice.

Just as cytological assessment is prejudiced so is colposcopic assessment. The squamocolumnar junction is less likely to be fully visualized and the ectocervix easily traumatized. Biopsy material, which should always include the squamocolumnar junction will nearly always require some type of conization; the shrinkage of the cervix can render this difficult. Major degrees of cytological abnormality in elderly women should always raise the possibility of invasive disease, usually cervical but occasionally even endometrial cancer might present because of abnormalities on a smear. Such smear results must be thoroughly investigated with early recourse to a large biopsy and endometrial assessment. As future fertility is not an issue in these women fear of anatomical distortion of the residual cervix should not deter the clinician from acquiring large biopsies.

9.16   Adenocarcinoma-in-situ (CGIN or AIS)

In this condition preinvasive disease affects the glandular epithelium of the endocervical canal. It is much less common than squamous CIN and although it can occur in isolation the majority of cases are found to have CIN in association with it. About one fifth of all cervical neoplasms have glandular elements yet the ratio of preinvasive glandular abnormalities to squamous is approximately 100 to 1 suggesting that not all glandular cancers have a preinvasive phase or it is underrecognized. The latter is probably true as smears are less effective at finding the condition and as the lesions are often small and sited at the base of glandular crypts they are less amenable to colposcopic recognition and therefore directed biopsy.

The treatment for AIS is somewhat controversial. Some authorities believe that, as the whole endocervical canal might be involved hysterectomy is necessary. Others however feel that this is overtreatment and cone biopsy is curative in the majority and should at least be the first line of management in the younger and nulliparous woman who has confirmed AIS.

9.17   Other gynæcological premalignancy

Dysplastic change can affect the epithelium of other areas of the lower genital tract either as a unifocal lesion or as multifocal disease. The squamous epithelium of the vagina may be involved (VaIN) either alone or in combination with extension of cervical disease. The vaginal epithelium might also be involved by extension from a vulval (VIN) preinvasive lesion.

Far less is known regarding VaIN and VIN as they are less common than CIN and are not the subject of a screening strategy. VIN is usually recognized as a result of symptoms (pruritus, soreness etc.) and changes visible to the naked eye such as hyperkeratosis (often referred to as leukoplakia), erythema and irregular surface contour. Diagnosis and management of these lesions requires considerable expertise.

9.18   Summary

Managing abnormal smears requires a basic understanding of the process of cervical dysplasia and considerable skill in understanding and managing the anxieties that the problem generates in women. A clear and lucid management plan should complement basic and understandable explanations to women. The screening programme is here to stay, at least for the foreseeable future. The need to manage the inevitable abnormalities that it generates must therefore be seen as a basic medical skill.

Learning points

CIN 1, CIN 2 and CIN 3 are dysplastic, preinvasive and asymptomatic abnormalities of squamous epithelium on the cervix

CIN 3, if untreated, may progress to cancer in 30–40% of women

The evidence the CIN 1 becomes CIN 3 is poor

The cervical screening strategy aims to detect and treat preinvasive disease and thus reduce the incidence of cancer

Most cases of CIN can be treated simply and effectively by outpatient excision or ablation

Careful counselling and explanations are required at all stages of the screening and treatment process

Other types of preinvasive disease include AIS, VaIN and VIN. These are much less common and require specialist attention

Further reading

Kolstad, P. and Stafl, A. (1982) *Atlas of Colposcopy*, 3rd edn. Churchill Livingstone, London.

Luesley, D.M. (1992) Advances in colposcopy and management of cervical intra-epithelial neoplasia. *Curr. Opin. Obstet. Gynecol.*, 4, 102–8.

Shafi, M.I. and Luesley, D.M. (1995) Management of low grade lesions-Follow-up or treat? *Clin. Obstet. Gynaecol.*, 9, 1.

Shafi, M., Luesley, D.M. and Jordan, J.A. (1992) Mild cervical cytological abnormalities. *Br. Med. J.*, 305, 1040.

10 Gynæcological cancers

Charles Redman

10.1 What is gynæcological cancer?

Gynaecological cancer includes all those malignancies that arise from the female genital tract. In some situations the genital tract, particularly the ovary, may be the site of metastases from distant sites such as the breast or stomach. Although important to note, secondary tumour spread to the genital tract will not be considered further in this chapter.

At each site within the genital tract a number of different malignant tumours can arise and the principal types are summarized below.

Site	Type	%	Median age of presentation
Vulva	Squamous	85	75
	Melanoma	5	
Vagina	Squamous	90	75
Cervix	Squamous	90	46
	Adenocarcinoma	10	
Uterus	Adenocarcinoma		60
Ovary	Adenocarcinoma		50

10.2 How common is gynæcological cancer?

In the United Kingdom over 13 000 new cases of gynaecological cancer occur each year accounting for 1 in 7 of all cancers in women (Figure

10.1). Approximately 8000 women die annually from gynaecological cancer, of which, ovarian cancer is the most common accounting for over 4000 cases. These figures are of importance in that they detail the size of the problem and the associated implications for health resource allocation. Perhaps of greater interest is how these data have changed over the years. Some cancers have become more common. The incidence and mortality rates for ovarian cancer have increased since the beginning of the century (Figure 10.2). Some of this increase

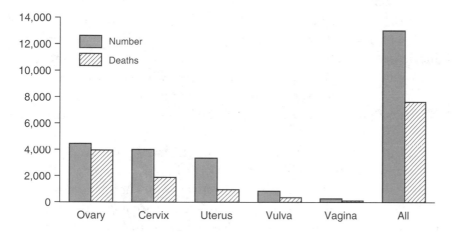

Figure 10.1 The annual incidence and deaths from gynaecological cancer. Source: Office of Population Censuses and Surveys (1985, 1988).

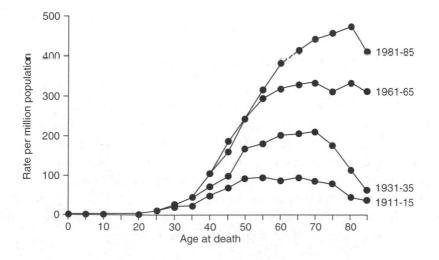

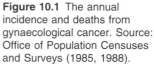

Figure 10.2 Death rates per million population for ovarian cancer – England and Wales 1911–85. Source: Office of Population Censuses and Surveys (1985).

may be attributable to more accurate diagnosis, especially in the older age groups, but the remainder must represent a true increase in incidence. Overall, there has been a doubling in the mortality rate over the last 70 years, which probably reflects the change in the nation's demography and strongly emphasizes the important association between ovarian cancer and age. There have, however, been decreases in both the incidence and mortality for uterine and cervical cancer although there has been a huge increase in the incidence of cervical precancer. Some of the fall noted for cervical cancer may be attributable to the cervical screening programme, but, only in part, as the mortality from cervical cancer was falling well before the screening programme was introduced.

These are overall changes but there are also some important changes within each disease type. For example, there has been an increase in incidence and mortality from cervical cancer in younger women, perhaps reflecting a change in sexual mores, whereas there has been a small decrease in ovarian cancer occurring in women aged under 55 years, possibly resulting from the use of the oral contraceptive pill.

The incidence of gynaecological cancers varies with age. In general terms most of the common cancers occur in older women (>40 years). Those seen more frequently in younger women include cancer of the cervix and also the rarer forms of gynaecological malignancy (for example malignant germ cell tumours of the ovary) are proportionately more common. This has to be borne in mind as these rarer malignancies often require a completely different therapeutic approach.

10.3   What causes gynæcological cancer?

The likelihood of a cancer developing within an individual appears to be a function of susceptibility on the one hand, be that hereditary or acquired (immunosuppression), and involvement of other factors such as exposure to carcinogens. Individual patients who have one sort of cancer can be more at risk of developing one at another site. In this way vulval and cervical cancer can be associated and there is a significant association between primary cancer of the breast, ovary and endometrium. Some of the recognized risk factors are:

	Factor	Ovary	Uterus	Cervix	Vulva
Demographic	Age	45+	60+	40+	65+
	Social Class	Higher	Higher	Lower	Lower
Genetic	Family history	Yes	Yes	?	No
	Breast cancer	Yes	Yes	No	No
Ovulatory	Late menopause	Yes	Yes	No	No
	Nulliparity	Yes	Yes	No	No
Other	Pelvic irradiation	Yes	No	No	Yes
	Obesity	?	Yes	No	Yes
	Smoking	No	No	Yes	Yes
	COC Pill	No	No	?	?

In certain types of gynaecological cancer there are undoubtedly hereditary factors. It is estimated that 5–10% of ovarian cancers may be hereditary. This is particularly so if the disease occurs before the age of 50. The overall lifetime risk for developing ovarian cancer is 1 in 120 but this rises to 1 in 40 if a woman has one affected first degree relative (e.g. mother/sister) diagnosed with the disease before the age of 50 and 1 in 3 if there are two affected first degree relatives similarly affected. Ovarian and endometrial cancer can occur more commonly in families in which other members have breast or bowel cancer. These facts emphasize the importance of taking a full family history when a woman has gynaecological cancer. Similarly there may be scope for family counselling.

For certain cancers environmental factors appear to be foremost. For example, sexual activity is a recognized risk factor for cervical cancer. Cervical cancer rarely occurs in virgins and the relative risk is seen to be higher in women who have started intercourse at an early age and or have had multiple partners. The implication is that early exposure of the cervix to some potential mutagen, such as certain oncogenic viruses or indeed semen itself, predisposes to malignant change. The premise that sexual activity (itself a function of social culture) has an important association with cervical cancer is supported by the observation that the incidence of cervical cancer can differ markedly between different populations. Thus cervical cancer is noted to be relatively uncommon in Israeli Jewish women but more so in Puerto Ricans. Nevertheless other factors must be involved as only a proportion of 'at risk' individuals do in fact develop cervical cancer, the majority of cases occurring in women who might not fit this type of model. Smoking is a recognized co-factor and although the mechanism by which this may function is not fully understood, it has been shown that local cellular immunity is influenced by cigarette smoke metabolites. The body's immune system may have an important role in preventing malignancy from developing as certain cancers, cervical cancer for example, may occur more commonly in patients who are immunocompromised.

Increased exposure to oestrogens of one sort or another predisposes to endometrial hyperplasia which can be a precursor for endometrial cancer. This can arise endogenously, as in obese women who have higher levels of oestrogens as a consequence of an increased amount of androstenedione being converted into oestrone by adipose tissue. Other endogenous instances include high oestrogen levels caused by oestrogen-secreting ovarian tumours, or in women who have anovulatory menstrual cycles, such as can occur in polycystic disease of the ovary. But exogenous oestrogen can also be important as was noted in the 1970s when endometrial cancer was noted to occur more commonly in postmenopausal women given continual oestrogen (without progestagen) as hormonal replacement therapy.

Ovulatory and reproductive history are also risk factors. Women with cervical cancer tend to be parous whereas higher rates of endometrial and ovarian cancer occur among women who are nulliparous, and that this risk decreases with the number of pregnancies. A raised risk of endometrial and ovarian cancer has also been reported in women who have a late menopause and who have a history of infertility. Could it be that ovulation itself predisposes to ovarian cancer? It had been suggested that cancer of the ovary can occur more commonly on the right ovary which is said to ovulate more often than the left and the oral contraceptive pill reduces the incidence of ovarian cancer because it suppresses ovulation.

Little is known concerning the aetiology of vulval cancer. A viral aetiology involving oncogenic human papillomaviruses has been suggested as have pre-existing skin disorders such as lichen sclerosus et atrophicus. This condition, once termed a vulval dystrophy was, until recently, not thought to be associated with cancer unless cellular atypia was also present. Several authors, however, have now commented on the frequency with which these skin changes are seen in association with invasive cancer. This might purely represent an association and not a cause and effect but it should certainly lead to close and pro-tracted surveillance of women found to have conditions such as lichen sclerosus et atrophicus.

10.4   Prognosis

This varies greatly. The overall 5-year survival ranges from almost 100% for early invasive cervical cancer to less than 5% for advanced ovarian cancer. In 1989 ovarian cancer killed 4411 women in the UK, more deaths than from cancer of the cervix and uterus combined. One has to give consideration as to why this should be. Prognosis is a complex function of tumour type, the degree of tumour spread (stage), and the ability to offer effective treatment.

In general terms the earlier the diagnosis the better the outcome will be because there is less likelihood that the disease has spread to adjacent or distal parts of the body. Gynaecological cancers can be staged according to predetermined and widely agreed guidelines, such as the TNM (Tumour, Nodal, Metastases) or the FIGO (Federa-tion Internationale de Gynecologie et Obstetriques) classification.

Percentage of ovarian cancer patients surviving 5 years (FIGO staging)

Stage	Description	Percentage surviving 5 years
I	Growth confined to ovaries	70
II	Growth confined to pelvis	45
III	Growth confined to abdominal cavity	17
IV	Distant disease (also intra-hepatic)	5
Overall		25

There is a clear association between stage and prognosis. For more lengthy descriptions of tumour stage the reader is referred to texts on oncology.

Survival can therefore be related to the timing of diagnosis. At one extreme, cervical cancer can be detected prior to clinical presentation as a result of the cervical screening programme and there is no doubt that patients with early cervical cancer, when it is confined to the cervix, will fare better than when it is more advanced. Similarly cancers that present early with symptoms will have a better prognosis. Endometrial cancer can cause bleeding, postmenopausal bleeding in particular, and this will often lead the patient to seek medical advice. Ovarian cancer, on the other hand, arises deep within the abdomen. Consequently three quarters of ovarian cancer patients present with advanced disease (FIGO Stages III–IV) whereas a similar number of patients with endometrial cancer will present with Stage I disease, when the cancer is confined to the uterus.

10.5   How does gynæcological cancer present?

Gynaecological cancer can occasionally be detected in asymptomatic patients. A proportion of cervical cancers (perhaps 10%) will be detected as a result of cervical screening, which can, very occasionally, also lead to the detection of other gynaecological cancers. Others may be detected serendipitously at the time of operation or in the course of investigation for a completely unrelated non-gynaecological reason, and occasionally a previously unsuspected cancer will be found in a hysterectomy specimen! It is therefore considered prudent not to perform a hysterectomy unless a cervical smear has been performed within 3 years and, if there has not been prior endometrial sampling, to open the hysterectomy specimen at operation in case there is an unsuspected endometrial cancer.

Cervical screening apart, early presymptomatic detection of gynaecological cancer requires close surveillance of identified high risk groups. Examples of such risk groups would be women who have one or two first degree relatives with ovarian cancer, or, with regard to vulval cancer, women who have been identified as having vulval intraepithelial neoplasia (VIN).

Symptoms of gynæcological cancer

Vulva	Lump, ulcer, soreness (Figure 10.3)
Cervix	Irregular vaginal bleeding, offensive discharge, deep visceral pain, nerve root pain
Endometrium (Uterus)	Postmenopausal bleeding, irregular heavy periods (Figure 10.4)
Ovary	Abdominal discomfort, Abdominal swelling (Figure 10.5)

Most gynaecological cancer is symptomatic at the time of presentation. The principal symptoms are summarized below.

An important point to make is that many of these symptoms are common and non-specific and that cancer will not be diagnosed unless a thorough examination and appropriate assessment is performed. Sadly the disease may present long before the diagnosis is made. Ovarian cancer, for example, is rarely symptomatic but its frequently vague and non-specific symptoms of mild gastrointestinal disturbance are often ignored by the patient and not uncommonly by the doctor when she initially seeks advice. Therefore one must have a high degree of vigilance particularly in women over 40 years of age. In this age group abnormal menstrual bleeding, be it in timing or amount, should be regarded with suspicion and requires a gynaecological examination. Endometrial sampling should also be seriously considered. Postmenopausal bleeding is cancer till proven otherwise. Postcoital bleeding also should arouse the suspicion of cervical cancer and at the very least the cervix should be visualized and a smear taken.

Unfortunately late diagnosis can occur because the patient presents late despite prolonged symptoms. Self-denial and delay in seeking help usually result from fear, not only of the disease but of its treatment and, in certain countries, its cost. Genital cancers also, quite wrongly, are felt to represent something unhygienic and as such generate embarrassment in addition to fear. This applies to vulval cancer in particular.

Figure 10.3 Vulval cancer. ©
Medical Illustration, North
Staffordshire Hospital.
Reproduced with permission.

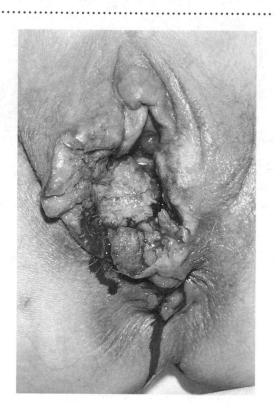

Figure 10.4 Uterine cancer.

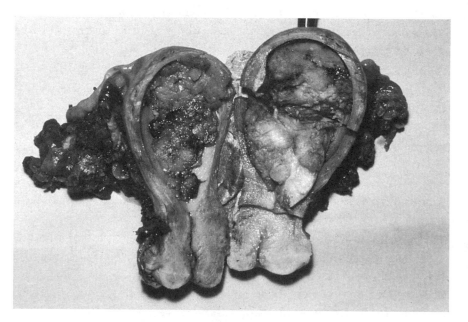

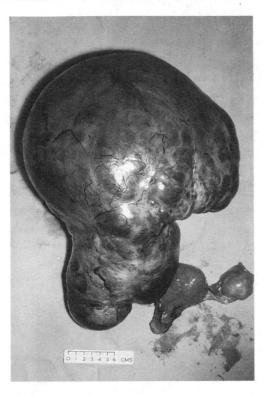

Figure 10.5 Ovarian cancer.

10.6 Investigation of patients with suspected cancer

In any situation when cancer is suspected the appropriate investigations must be done as expeditiously as possible. Such investigations should address several important questions:

10.6.1 What is the diagnosis?

The diagnosis of cancer must be supported by histological proof. In vulval, vaginal and cervical cancer this is usually obtained by directed biopsy. In endometrial cancer the endometrium can be assessed under direct vision using a hysteroscope either in outpatients or in theatre as an inpatient and a directed endometrial sample taken. More commonly, though, endometrial sampling is performed blind either by formal dilatation and curettage or using an outpatient sampling technique. Thus it is feasible, though not always appropriate, to obtain histological confirmation at the first outpatient consultation in most of the common gynaecological malignancies. This is not the case for ovarian cancer, where laparotomy remains the principal procedure to obtain histological confirmation.

10.6.2　Is there spread?

The extent of spread has profound prognostic and therapeutic implications. For example the detection of hepatic involvement in a patient with endometrial cancer could be a significant contraindication to hysterectomy, even if it were technically feasible. In addition to clinical examination the detection of spread can be assessed in one of two ways. First by certain non-invasive investigations which can usually be arranged and performed in conjunction with the diagnostic procedures. The actual tests depend on the type of cancer and its mode of spread.

Commonly performed non-invasive investigations of tumour spread

Investigation	Looking for	Ovary	Uterus	Cervix	Vulva
Chest X-ray	Chest metastases	Yes	Yes	Yes	Yes
	Pleural effusion	Yes	Yes	Yes	No
Ultrasound	Liver metastases	Yes	Yes	No	No
	Ascites	Yes	Yes	No	No
	Ureteric involvement	No	No	Yes	No
CT Scan	Retroperitoneal nodes	No	No	Yes	No
IVU	Ureteric involvement	No	No	Yes	No
Biochemistry	Liver/renal problems	Yes	Yes	No	

A chest radiograph is always an important test in this situation as the detection of distal spread can have a major influence on subsequent management. Similarly the possibility of hepatic or ureteric involvement can easily be detected. Clinical examination combined with non-invasive investigations are referred to as a clinical staging or assessment.

Second, there is surgical assessment where the extent of the disease involvement is assessed operatively and by pathological assessment of the surgical specimens providing data that could not be obtained in any other way. This is particularly relevant in ovarian cancer in which formal laparotomy is the most sensitive and accurate test for detection of disease spread. Surgical assessment also forms a pivotal part in the staging of endometrial cancer.

Staging forms part of assessing how advanced a cancer is and, depending on the disease, can be either clinical or surgical. This is determined by the accessibility of the disease and how it is managed. The external cancers, such as vulval and cervical cancer, are staged clinically whereas the more inaccessible endometrial and ovarian cancers are staged surgically. Whether all cancers, cervix in particular, should be assessed by a staging operation is a point of current debate. Surgical staging certainly provides more and more accurate information regarding tumour spread but whether or not this additional information (such as knowing whether the lymph nodes have cancer in them) translates into better treatment and improved survival remains doubtful. In the UK surgical staging for cancer of the cervix is not generally accepted thus the stage of cervix cancer does not take into account the status of the regional lymph nodes despite the fact that these nodes are a known site of metastases. A diagrammatic representation of the staging of cervical cancer is given in Figure 10.6.

10.6.3 Operative risk?

Treatment decisions are not only influenced by the type and extent of disease but also the general health of the patient. This is well exemplified in early cervical cancer where younger and fitter women may have surgery whereas older fatter women may be offered radiotherapy. Many patients with gynaecological cancer will require surgery so it is good practice to assess surgical fitness straight away and organize the appropriate investigations. Age alone should not be considered as a contraindication to surgical management. Many elderly women are either fit or can be made fit to tolerate radical surgery and this should always be considered, particularly if such an approach might result in cure.

10.7 Making a diagnosis and counselling the woman

Most patients want to know what is going on and to receive accurate information. Inaccuracies, inconsistent and conflicting facts only heighten distress and reduce confidence. In an ideal world a calm patient will present early to a doctor who can instantly make the right

Figure 10.6 Clinical staging of cervical cancer.

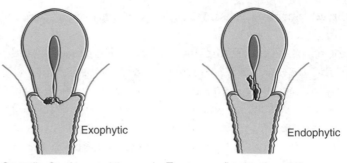

Exophytic   Endophytic

Stage Ib: Carcinoma of the cervix. Tumour confined to the cervix

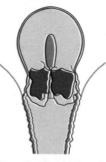

Stage Ib: 'Barrel' tumour expanding but still confined to the cervix

Stage IIa: Tumour extended on to vagina but not lower third

Stage IIb: Tumour extends to parametrium but not as far as the pelvic sidewall

Stage IIIa: Tumour extends to lower third of the vagina

Stage IIIb: Tumour extends to parametrium with no tumour-free space between cervix and pelvic sidewall

diagnosis and easily arrange the appropriate investigations and treatment which would result in cure. We know this is not the case for a variety of reasons. First, the diagnosis may not be obvious and cancer is only one of a number of possible options. The problem here is to take appropriate and timely action without causing undue concern. Patients should not be told they have cancer unless the diagnosis is certain or beyond reasonable doubt. Second, it is inevitable that a number of professionals will be involved, emphasizing the need for good communications; this includes an understanding of what events are likely to follow a consultation or procedure. If a patient is told that she may have cancer she will want to know what happens next. Finally, few patients are calm; many patients are anxious when they see a doctor for whatever reason but in this context very much more so. As a result only a small part of what is said will be fully understood and much might be misunderstood. Information must therefore be simple, short and preferably written down and it is always desirable that the patient have a friend or relative with her. Information sheets and booklets can have great value in allowing the patient the opportunity to go through things at her own pace although such information should be supported by allowing patients ready access to medical or nursing staff to answer other questions and deal with problems that can arise.

10.8 Anxiety and cancer

There is a danger that in these circumstances management focuses more on the cancer than the patient herself, and her family. There can be few things as profoundly distressing as to be told that you or a loved one has cancer. The distress engendered arises from many sources, be that a fear of dying, pain or mutilation or simply a fear of the unknown resulting in a feeling of utter helplessness.

Help can come from a number of sources. First, from medical and nursing staff, both within the hospital and the community. Medically the provision of clear and consistent information is an important priority. Many patients and their families do value the opportunity of talking to medical staff. Nevertheless support from other sources is of immense importance as patients will require different types of support in different circumstances and, most of all, time. There is now a greater awareness of the need for professional cancer patient support, including input from cancer counsellors and psychologists who can help the patient come to terms with the situation. Another very important form of support is that which can come from other patients in self-help groups.

10.9   Defining the objectives of treatment

The therapeutic effectiveness of a cancer treatment is a balance between its cancer destroying activity and the toxicity to the patient. Sadly, increased activity is often associated with increased toxicity but this may be justifiable if the end results can justify the means. Consequently an important initial clinical decision is whether treatment is intended or likely to be either curative or palliative. In deciding which, the clinician must address two important questions. First, is the cancer curable (this is largely determined by disease factors such as what the cancer is and whether it is advanced or not) and secondly what sort of treatment is most applicable for the patient in question?

A good example of a curative treatment being considered on disease and patient grounds is when surgical exenteration (removal of the pelvic viscera, including the bladder and rectum) is considered for the management of recurrent malignant disease, often cancer of the cervix. In this uncommon situation the operation is virtually always performed with curative intent because the surgical hazards and the extent of the surgery can only be justified on those grounds. Extensive investigations are performed to exclude metastatic disease and to confirm that the disease is curable (disease factors) and, even if it is, only patients suitable in terms of general heath, mental attitude and age would be considered (patient factors).

In practice, however, the distinction between palliation and radical therapy (i.e. with curative intent) can be blurred. Over 75% of patients with ovarian cancer will have disease that cannot be removed completely at initial surgery and most of these patients will die from ovarian cancer. In this context one could argue that treatment is primarily palliative, although many patients will in fact be asymptomatic after their initial operation. Nevertheless radical (and often toxic) treatment is given with curative intent in order to try and maximize the chances of a cure for that individual patient.

In other situations the extent of the disease may be such, or the general condition of the patient so poor that active therapy would only be considered if there were definite advantages to the patient's quality of life.

10.10   Management methods in cancer

Cancer therapies aim to reduce the amount of cancer either by removing it surgically, by using cytotoxic treatments such as radiotherapy or chemotherapy, or slowing its growth, such as the use of progestagens in cancer of the endometrium. These therapies can be used singly or in combination.

Surgery alone has an important role in the radical management of early gynaecological malignancy as is the case in cancers of the vulva (radical vulvectomy), cervix (radical or 'Wertheim's' hysterectomy), endometrium and ovary in which long-term survival is achievable without recourse to additional therapies. In certain circumstances, although surgery is the principal or primary radical treatment, further treatments may be given. As an example, following surgery for early cervical cancer some women are advised to have radiotherapy, usually because the pelvic lymph nodes removed at operation are found to contain metastases (this will be the case in about 20% of such cases). This type of therapy is called adjunctive therapy, i.e. additional treatment given to eradicate disease that the clinician suspects may be present. On occasions some patients may receive an additional treatment prior to the main primary radical treatment to try and improve results. This is termed neoadjuvant therapy (a sort of enabling therapy) and examples of this include the use of preoperative radiotherapy in cancer of the endometrium and vulva and the use of chemotherapy prior to radiotherapy or surgery in cancer of the cervix. A general rule, and emerging philosophy, in the treatment of 'Early' gynæcological cancers is to maintain the expected high cure rates but to reduce the morbidity and toxicities associated with treatment.

In advanced malignancy the effectiveness of localized treatment diminishes as the likelihood of regional and distant spread increases. For example, only 1% of patients with early (Stage I) cervical cancer will have affected para-aortic lymph nodes, whereas when the disease has spread from the cervix (Stage II) this will rise to 30%. In these circumstances there is a need to use treatments that can be cytotoxic over wide areas if not systemically. Therefore radiotherapy, chemotherapy and hormonal therapy assume far greater importance. The radical role of surgery is diminished and, if used, forms part of a combined modality cytoreductive approach. Cytoreductive surgery in advanced ovarian cancer will not reduce the need for postoperative chemotherapy but it might enable the chemotherapy to be more effective and as such is perhaps an example of surgery as a neoadjuvant therapy.

10.11 Palliation

In many patients either at initial presentation or at a subsequent relapse it is recognized that the chances of cure are remote and that radical treatment, with its implied greater toxicity, would be inappropriate. In these circumstances the aim is to relieve symptoms with minimal toxicity. This would be termed a palliative approach. Palliation can utilize any or none of the different types of therapy used radically. Surgery may be employed, for example, to reduce distressing symptoms from intestinal obstruction or faecal fistulae. Radiotherapy can provide quick and valuable resolution of discomfort from ovarian cancer masses or to control vaginal bleeding. Chemotherapy may be used to reduce pain, in cervical cancer for example. There are, however, many other aspects of treatment that have a major role, such as pain relief, or relief of nausea.

Learning points

Gynæcological cancers account for a significant amount of morbidity and mortality in women of all ages
The types of cancer vary with the majority occurring in women over the age of 40
The outcomes depend on a relationship between tumour type, site stage and the effectiveness of available therapies.
Good management relies on prompt recognition, appropriate investigation with regard to diagnosis, extent and general fitness to tolerate treatment.

Learning points (*Continued*)

At no time should the whole patient, and indeed family unit be forgotten

Counselling and communications skills are all important oncological skills.

The objectives should be clearly defined if possible so that a correct balance between radicality and morbidity can be made

Palliative treatment is as important as treatment intended to cure

All available treatment modalities are employed in the management of gynæcological malignancies

Further reading

Shepherd, J.H. and Monaghan, J.M. (eds) (1990) *Clinical Gynaecological Oncology*, 2nd edn. Blackwell Scientific Publications, Oxford.
Ovarian Cancer, Cancer Research Campaign Factsheets 17. 1-1, 1991.

11 Uterovaginal prolapse

Sarah Creighton and Frank Lawton

11.1 Introduction

Genital prolapse is a very common condition and more so in the elderly. Prolapse and prolapse-related problems account for nearly a quarter of women waiting for routine gynæcological surgery. The condition is rarely life threatening but can cause considerable discomfort and distress. A basic understanding of prolapse relies heavily on a knowledge of the anatomy of the female pelvic floor and on the processes both physiological and pathological that lead to malfunction of the pelvic musculature and ligaments. Only with this knowledge can an adequate explanation be given to women with prolapse and only with this knowledge can sound management be planned.

11.2 What is prolapse?

In the healthy woman, the pelvic floor acts as a hammock or sling for the pelvic organs. A prolapse can be described as a type of hernia resulting from a weakness in the musculofibrous tissue that constitutes the pelvic floor. These herniae present as bulges or masses in the vagina that may just cause a sensation of a mass and/or cause a disturbance of function of the prolapsing structures such as the bladder or rectum. The type of problems can best be appreciated once the structure and function of the pelvic floor are understood.

11.3 What supports the pelvic organs?

11.3.1 Muscles (Figure 11.1)

A pair of symmetrical striated muscles constitute the major part of the pelvic floor. As a group these muscles are termed the levator ani muscles. These muscles fuse in the midline but are perforated by openings for the urethra, the vagina and the rectum. The levator ani can be further separated into three parts although it functions as one muscle. The most anterior part is termed the pubococcygeus and runs from the back of the pubic symphysis to insert into the tip of the coccyx. This part forms slings around the urethra forming a part of the urinary continence mechanism. In addition, fibres pass around the vagina and the rectum forming part of the external anal sphincter. The middle section is called the iliococcygeus, this overlaps with pubococcygeus and inserts below it on the coccyx and also fuses in the midline to form the anococcygeal raphe. Finally the most posterior part is the coccygeus. This smaller section runs from the ischial spine and inserts into the sacrum and coccyx. All three muscles are in continuity with each other. These muscles receive their nerve supply from the pudendal nerve and also from direct branches of the sacral roots (S3 and S4.)

The pelvic floor muscles are striated muscle but act differently from striated muscle elsewhere in the body such as limb muscles. The pelvic floor is constantly in an active state, even in sleep, yet has to adapt to situations of excess strain such as coughing and to relax in order to

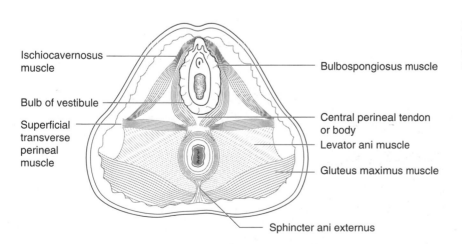

Figure 11.1 Muscles of the pelvic floor and vulva. Adapted from *Memorix Gynaecology*, Chapman & Hall, London, 1996.

Ischiocavernosus muscle

Bulbospongiosus muscle

Bulb of vestibule

Central perineal tendon or body

Superficial transverse perineal muscle

Levator ani muscle

Gluteus maximus muscle

Sphincter ani externus

allow defecation and micturition. To achieve this it is supplied by a mixture of fast and slow twitch fibres. The slow twitch fibres provide baseline tone and constant support for long periods. The fast twitch fibres are recruited for extra support during sudden onset of strenuous activity such as exercise or coughing.

11.3.2   Ligaments (Figure 11.2)

The pelvic floor muscle is enveloped in a sheet of fascia which is attached to the pelvic brim and covers the entire pelvic floor. In areas it is condensed to provide strong supporting ligaments. These include the uterosacral ligaments and the cardinal (or transverse cervical) ligaments. The uterosacral ligaments are the most prominent and easily identifiable running from the back of the uterus and cervix to the sacrum. The cardinal ligaments run laterally from the cervix to the pelvic side walls. The anterior fascia is also condensed in the pubourethral ligament which anchors the urethra to the back of the pubis and the pubovesical fascia which runs from the pubis to the back of the bladder.

The bladder, rectum, upper vagina and uterus therefore lie on top of the pelvic floor and thus depend on a combination of muscular and fascial support. As with any mechanical system, weakness is inherent if the supporting structures are perforated as is the case with the pelvic floor. The vagina is a focal point for potential weakness, not only because of its central pelvic position but also because of physiological processes that damage the integrity of it and its supports.

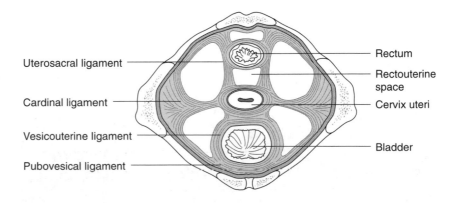

Figure 11.2 Uterine supports. Adapted from *Memorix Gynaecology*, Chapman & Hall, London, 1996.

Uterosacral ligament

Cardinal ligament

Vesicouterine ligament

Pubovesical ligament

Rectum

Rectouterine space

Cervix uteri

Bladder

11.4 What factors weaken these supports?

Factors contributing to uterovaginal prolapse

Childbirth
Abnormal collagen
Raised intra-abdominal pressure
Hormonal status
Previous pelvic surgery

11.4.1 Childbirth

The most important factor in the aetiology of prolapse is childbirth. Vaginal delivery causes damage by stretching and mechanical distortion of the relatively inelastic ligaments and also through a mechanism of partial denervation of the muscles. This results in a reduction of overall mechanical integrity and eventual weakness of support. Women who have not had vaginal deliveries appear to have a lower incidence of uterovaginal prolapse suggesting that most damage is done at the time of vaginal delivery and a much lesser degree of damage is sustained during pregnancy. Denervation damage is aggravated by having a long second stage and a large baby. Having an episiotomy does not prevent damage to the pelvic floor. Ligaments and fascia also undergo considerable stretching during delivery and these may also be irreparably damaged.

Prolapse is therefore most common in multiparous women. It may cause problems immediately after delivery but more commonly will present later in life.

11.4.2 Collagen

The majority of women with prolapse are multiparous, however some nulliparous women will also develop prolapse. Conversely many women who have had several children do not develop prolapse. It has been suggested that there is a subtle difference in the collagen of women who develop prolapse. Collagen is a major constituent of the fibrous structure making up the pelvic floor and an underlying collagen weakness may predispose women to prolapse with few or no children. Women with Ehlers–Danlos syndrome (a collagen disorder) have a much higher incidence of prolapse. Conversely, negroid women have a much lower incidence of prolapse and this may also be due to a difference in collagen.

11.4.3   Raised intra-abdominal pressure

Any condition that causes raised intra-abdominal pressure over prolonged periods of time may cause pelvic floor weakness. A chronic cough due to smoking will certainly cause this. Other factors include chronic constipation leading to prolonged straining and also gross obesity. Unfortunately, once prolapse is present, giving up smoking or losing weight will not cure the condition although it may make any surgical treatment easier to perform and less hazardous to the patient. A pelvic mass such as a fibroid uterus or ovarian carcinoma may also cause increased intra-abdominal pressure and should be excluded on vaginal examination.

11.4.4   Hormonal status

Prolapse is more common after the menopause. Low oestrogen levels may result in weakening and atrophy of the muscular and ligamentous supports and contribute to the development of prolapse. There is, however, no evidence that giving hormone replacement therapy reduces the incidence of prolapse.

11.4.5   Previous pelvic surgery

Previous pelvic floor surgery weakens the pelvic supports. Total abdominal and vaginal hysterectomy may predispose to the development of prolapse as both of these operations cut across the supporting ligaments (uterosacral and cardinal ligaments). A colposuspension operation for stress incontinence lifts up the front wall of the vagina but stretches and weakens the supports posterior to the vagina allowing the back wall of the vagina to descend causing prolapse.

11.5   Types of prolapse

Any part of the vaginal wall may herniate through the vaginal opening (Figure 11.3). The type of prolapse depends upon the structure directly adjacent to the vaginal wall and thus involved in the prolapse.

Types of prolapse

Cystocele	A prolapse of the anterior vaginal wall and contains bladder (Figure 11.4a)
Rectocele	A prolapse of the posterior vaginal wall and contains rectum (Figure 11.4b)
Enterocele	A prolapse of the pouch of Douglas (the portion of vagina lying directly behind the uterus) and usually contains loops of small bowel. An enterocele may occur at the same time as a rectocele and be continuous with it (Figure 11.4c)
Uterine	The uterus can also descend through the vagina and the degree of uterine descent is described by the relative position of the cervix at the time of examination when the patient is asked to strain
First degree	The cervix is seen to descend into the vagina but not as far as the introitus (The entrance to the vagina) (Figure 11.5a)
Second degree	The cervix descends to the introitus or just beyond so that the uterine body is still within the pelvis (Figure 11.5b)
Third degree (Procidentia)	The cervix and uterine body have prolapsed beyond the introitus, i.e. fully on view between the patients thighs (Figure 11.5c)
Vault prolapse	The vagina prolapses in a woman who has already had a hysterectomy. The vagina becomes turned inside out, like a sock, and hangs outside the introitus

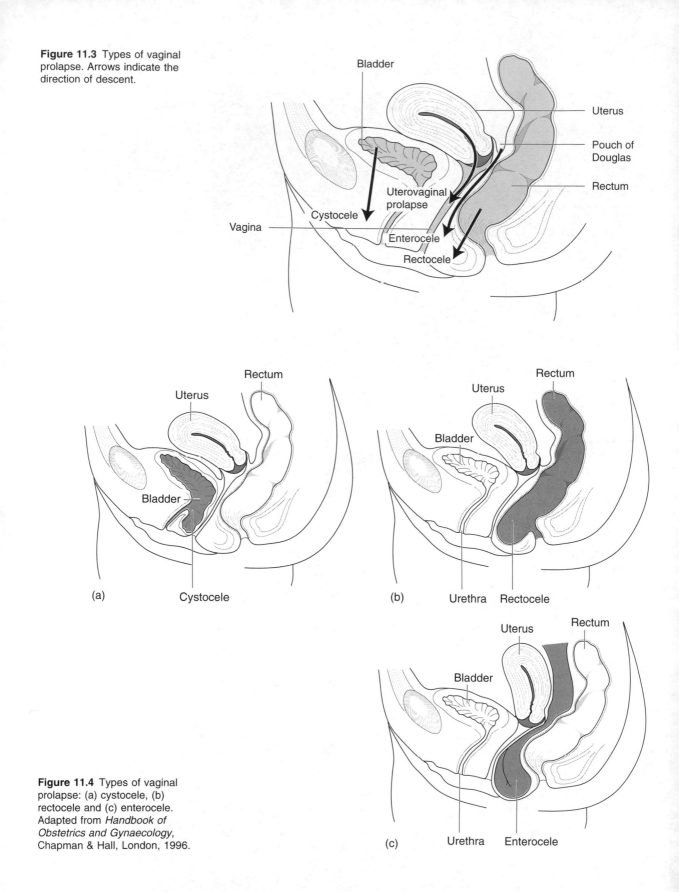

Figure 11.3 Types of vaginal prolapse. Arrows indicate the direction of descent.

Figure 11.4 Types of vaginal prolapse: (a) cystocele, (b) rectocele and (c) enterocele. Adapted from *Handbook of Obstetrics and Gynaecology*, Chapman & Hall, London, 1996.

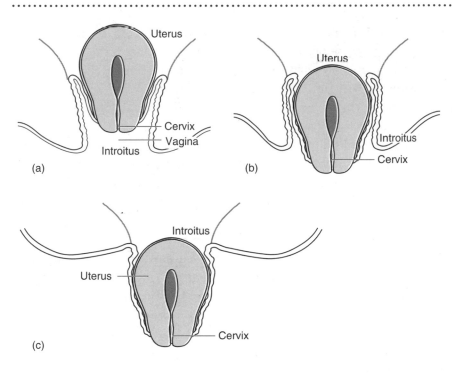

Figure 11.5 Diagrammatic representation of degrees of uterovaginal prolapse. (a) **First degree** – cervix descends into vagina. (b) **Second degree** – cervix prolapsed beyond introitus. (c) **Third degree** or **procidentia** – cervix and uterus prolapsed beyond introitus.

11.6 Symptoms of prolapse

The most common complaint of women with prolapse is that they notice a lump in the vagina, 'something coming down' or that 'something has dropped'. This may only be present after prolonged standing or walking or, if severe, may be present all the time. It may be associated with a 'dragging' feeling and lower abdominal and back pain. A woman with a cystocele may develop urinary symptoms. The altered anatomy may cause kinking of the urethra and/or a bladder pouch which in turn may cause incomplete bladder emptying and urinary stasis. This will predispose to infection which is classically recurrent. As a consequence such women will often given a history of repeated attacks of 'cystitis', i.e. frequency and dysuria.

Rectoceles are usually asymptomatic and as such pose no major problems. However, the symptoms when they do occur relate to problems with defecation or a sensation of a mass. A woman with a symptomatic rectocele classically complains of a feeling of incomplete bowel emptying. She may have to return to the toilet a second time to completely empty her bowel. If the prolapse is large, she may have to manually reduce it back into the vagina to defecate at all.

Problems with intercourse are not uncommon and may be due to alteration in the shape or tone of the vagina, an off putting physical appearance or worry as to the cause of the prolapse.

Procidentias lead to a drying of the cervical and vaginal mucosa. This combined with rubbing of the skin on underclothes can lead to chaffing and soreness during walking. It may become ulcerated and infected with associated discharge and bleeding. Although the appearances of a neglected procidentia can be quite alarming it is not associated with an increased risk of vulval or vaginal carcinoma.

If the uterus prolapses through the vagina it will stretch adjacent structures the most important of which are the ureters. If they are stretched down with the uterus they may become obstructed and eventually cause renal failure. Although uncommon, the possibility of renal compromise should always be considered in the elderly woman with a long-standing procidentia.

11.7   Association with urinary and faecal incontinence

Childbirth can also damage both urethral and anal sphincters resulting in impaired control of urinary and faecal continence.

Stress incontinence of urine is a loss of urine upon coughing, straining or any manoeuvre that increases intra-abdominal pressure. It is due to a weakness of the urethral sphincter mechanism. Stress incontinence is often associated with a cystocele although both can occur individually. Careful enquiry about urinary symptoms must be made while taking the history of a woman with prolapse. Urinary incontinence in women is also dealt with by gynaecologists and both conditions may need treatment. Faecal incontinence is embarrassing to the patient and she may not report it spontaneously. The mechanism of faecal continence is dependent on several structures and the levatores ani are an important component. Faecal incontinence, although uncommon, can occur at the same time as uterovaginal prolapse.

11.8   Explaining prolapse to women

Prolapse is usually explained to the woman as a weakness developing in the muscles of the pelvis. This has allowed the bladder, uterus (womb) or bowel to 'drop'. The patient herself may be anxious having noticed a 'lump' as many patients associate any kind of lump with cancer. The benign nature of prolapse must be emphasized. Most women will have heard of a hernia and as the mechanisms are similar it is a useful way of explaining the various weaknesses that can occur in the pelvic floor. It is important to avoid medical jargon such as procidentia or cystocele as these terms will confuse and cause anxiety.

The possible causes should be discussed and advice as to weight loss or stopping smoking may be appropriate.

11.9   Examining women who have prolapse

The diagnosis of prolapse is made on history and examination. Few investigations are needed. Clinical examination for prolapse includes a general, abdominal and pelvic examination. Predisposing factors such as obesity or chronic cough should be noted and if possible remedial steps taken. The abdominal examination should aim to exclude pelvic masses and should also pay close attention to known sites of herniation. Concurrent inguinal or incisional hernias may indicate a collagen weakness. The pelvic examination is somewhat different from that performed for other gynæcological complaints. First, the patient is examined in the dorsal position and a routine bimanual examination performed taking care to examine the introitus to look for signs of bulging vaginal mucosa and or the cervix. The patient should be asked to cough or strain, preferably with a full bladder to demonstrate any associated urinary incontinence. The perineum should be inspected carefully for signs of deficiency or old scarring. Finally any signs of faecal incontinence should be sought. This might not have been volunteered at the time of taking the history. A bivalve speculum is then passed to examine the upper vagina and cervix. This speculum is not appropriate for determining the degree or site of prolapse but does allow the cervix to be inspected and a cervical smear taken if necessary. Hyperkeratosis, drying or thickening of the cervical epithelium would indicate that this structure is regularly prolapsing beyond the introitus and in these circumstances a smear may not be satisfactory. After removal of the bivalve speculum a bimanual examination is performed to ascertain the size, mobility, position and tenderness of the uterus. The adnexae and pouch of Douglas are also palpated for signs of pelvic masses that may have provoked the prolapse. After the bimanual examination has been completed a finger is gently inserted into the rectum whilst keeping the original examining finger in the vagina. By this manoeuvre, and asking the patient to strain it is sometimes possible to differentiate between a rectocele and enterocele. The woman is then asked to change position to either the left lateral or Simms position (semi-prone). In this position a Simms speculum is passed. This allows a good view of the anterior vaginal wall and again by asking the patient to strain any degree of cystocele can be recorded. The amount of cervical descent is also seen in this position although it is usually necessary to retract the anterior vaginal wall to visualize the other vaginal walls if a large cystocele is present. The Simms speculum is slowly withdrawn to allow the posterior fornix to be visualized and again if a bulge is noted on straining this would indicate either an enterocele or a high rectocele. Continuing to withdraw the speculum allows the whole posterior vaginal wall to be assessed and thus any rectocele that may or may not be present. If surgery is not contemplated then this is an appropriate time to fit a suitable pessary (see below) after discussing the options with the patient.

A mid stream urine specimen must be taken for culture and sensitivity to exclude infection, particularly if there is associated urinary frequency. An elderly woman with a chronic procidentia may need serum urea and electrolyte measurements to assess renal function. If a woman has co-existing urinary incontinence, urodynamic tests may be required (Chapter 12).

A woman with faecal incontinence may need a surgical referral.

The only other investigations are those required to assess general fitness and potential suitability for a surgical procedure.

11.10   Treatment of prolapse

Factors to consider when planning treatment

The examination findings and degree of prolapse
Age and frailty of the patient
The wish of the patient to continue sexual activity
The potential for further child bearing
Associated bladder or bowel symptoms
The woman's wishes after informed discussion

Surgery is the most common and effective method of treating prolapse. Conservative methods may be appropriate in a few clinical situations. Patients may have prolapse of all of the pelvic organs or they may have predominantly one type of prolapse e.g. primarily a cystocele. The principle of treatment is to elevate the prolapsed organ, push it back to its original position and repair the weakness or defect in the pelvic floor. Surgical correction of prolapse may entail removal of the uterus and employing the residual uterine supports to strengthen the remaining pelvic floor. Conservative methods rely on physiotherapy or more commonly the insertion of a vaginal pessary. A pessary is a flexible yet rigid plastic support inserted into the vagina thus providing an artificial pelvic floor.

11.10.1   Conservative treatment

Conservative treatment consists of supporting the prolapse with a pessary inserted into the vagina. The most common type is a ring pessary. This is inserted into the vagina and rests on the back of the pubic shelf anteriorly and in the sacral hollow posteriorly. The cervix protrudes through the middle of the ring. The size is adjusted until a ring is found which stays in comfortably yet retains the prolapse. Occasionally a ring pessary may be difficult to retain and a shelf pessary used. This is sited in the vagina with the handle facing downwards to allow insertion and withdrawal (you should ask to see the various types of pessary in use in your department).

Pessaries may be uncomfortable and occasionally lead to ulceration of the vaginal walls, particularly in the posterior fornix. This can cause a discharge and bleeding and rarely squamous carcinoma has been described occurring in a pessary ulcer. Occasionally a fibrous skin bridge may grow over the pessary and cause it to be stuck in the vagina. This may happen if the woman neglects to have her pessary regularly inspected and changed. The patient must have a vaginal examination every six months to change the pessary and inspect the vagina. Pessaries are therefore reserved for those women too frail to undergo surgery, those reluctant for surgery or as a temporary measure in those awaiting surgical repair.

Pelvic floor exercises are of little benefit in treating prolapse but may improve associated stress incontinence. Other treatments such as local or systemic oestrogens are of little benefit.

11.10.2 Surgical treatment

There are many operations available and the choice depends on the type and degree of prolapse. Most of the operations are performed by operating through the vagina. The woman usually has a general anaesthetic but may occasionally have an epidural if unfit for general anaesthetic. For vaginal surgery, the patient has her legs in the lithotomy stirrups so access to the vagina is simple. A repair operation on the vagina was previously termed a 'colporrhaphy' although nowadays the simpler term of 'repair' is used. Thus an anterior colporrhaphy is an anterior repair and a posterior colporrhaphy a posterior repair. 'Repair' is also more easily understood by the patient as it simply and accurately expresses the concept of the surgery.

(a) Repair of a cystocele

A cystocele is treated by performing an anterior repair. A longitudinal incision is made on the vaginal wall and the vaginal skin separated from the underlying bladder. The bladder is then pushed upwards. Stitches are placed in the fascia either side of the bladder and urethra and the tissues brought together to push up and support the bladder. Any excess vaginal skin is trimmed away and the vaginal wall closed with a continuous stitch. An anterior repair is a good treatment for a cystocele. It is not, however, a good treatment for stress incontinence and if the patient has combined incontinence and prolapse, an abdominal procedure may be better.

(b) *Repair of a rectocele*

The principle of a posterior repair is identical to that of an anterior repair. A longitudinal incision is made in the posterior vaginal wall and the skin separated from the underlying rectum. The rectum is then pushed upwards and backwards and stitches placed in the levatores ani each side. The levatores ani are brought together in front of the rectum to lift and support it. Any excess vaginal tissue is excised and the vaginal skin closed.

(c) *Treatment of uterine prolapse:*
vaginal hysterectomy

If uterine prolapse is present, the best treatment is a vaginal hysterectomy. All cutting and stitching is done through the vagina. This means the procedure causes less pain and the patient can mobilize quickly. The advent of antibiotics, particularly metronidazole has made vaginal hysterectomy a far less morbid procedure than it used to be as sepsis due to anaerobes is now uncommon. At the time of vaginal hysterectomy the ovaries can be palpated to identify any masses and if normal are usually conserved. Care is taken after the uterus is removed to repair any potential defect to prevent later development of an enterocele. A vaginal hysterectomy can be combined with an anterior and or posterior repair if a cystocele or rectocele is present.

(d) *Manchester repair (Fothergill operation)*

This operation is performed less often nowadays although it was an important and less morbid alternative to vaginal hysterectomy previously. It has been suggested as appropriate for women with uterine prolapse who wish to retain their uterus – usually in order to retain fertility. It is a vaginal procedure and involves removal of the cervix and approximation of the cardinal ligaments from each side. It can cause both cervical stenosis and cervical incompetence and recurrence of the prolapse is common. For these reasons it is becoming less popular.

(e) *Repair of other types of prolapse*

Enterocele repair   An enterocele can be repaired using the same procedure as for a rectocele. The levator stitches may, however, need to be higher insider the vagina. If an enterocele sac is present it may need to be opened to push back the small bowel and then remove and stitch the sac. Care must be taken not to make the vagina too tight in women who are sexually active.

Repair of a vault prolapse   The type of operation chosen to repair a vault prolapse depends on whether the patient is sexually active. If she is sexually active, a colposacropexy is the most appropriate procedure. This is performed through an abdominal incision. In this procedure the top of the vaginal vault is stitched to a synthetic mesh. This in turn is stitched under minimal tension to the front of the sacrum. This will hold the vagina up without making it narrower. It is however a major procedure and not suitable for frail elderly women.

If the woman is frail and not sexually active, a vaginal procedure is more appropriate. It is effective, less invasive and has a quicker recovery time. A tight anterior and posterior repair may be performed to hold the prolapse or alternatively the vagina may be completely obliterated with a series of circular stitches (a Le Fort procedure).

11.11   Recovery

Most women make a quick recovery from vaginal procedures. Pain is not severe and mobilization can be early. Occasionally there is difficulty in passing urine in the immediate postoperative phase. This is a result of levator spasm and oedema of the tissues of the pelvic floor. The woman should be reassured that function will return spontaneously and temporary catheterization instituted. The use of prophylactic antibiotics has greatly reduced postoperative sepsis.

11.11.1   Postoperative complications

Immediate postoperative complications include haematoma formation at the vaginal vault. This may become infected and present with a temperature and or an offensive discharge. Late postoperative complications can include dyspareunia (pain on intercourse). This can be due to scarring or narrowing of the vagina after a repair. Recurrent prolapse is unfortunately common. Risk factors such as obesity and smoking must be emphasized to the woman before she leaves hospital. She should also try to avoid heavy lifting, straining or other exercise that will put excess pressure on the newly repaired pelvic floor.

11.12   Conclusion

Prolapse is a common condition. Symptoms produced may be minimal or severe and incapacitating. The principles behind both the causes and treatment of prolapse are simple to understand. Careful assessment of the woman with prolapse is important to determine the most effective and appropriate treatment method.

Learning points

Symptoms of prolapse need not be related to the degree of anatomical distortion

Childbearing is the major predisposing factor

Any factor such as obesity, chronic cough or constipation will provide excess strain on the pelvic floor and aggravate prolapse

Uterovaginal prolapse may cause pain, problems with micturition, problems with defecation and problems with intercourse

Prolapse may frighten women as they associate swellings and masses with cancer.

Surgery is the preferred method of management

Vaginal pessaries are an alternative in women either unsuitable or unwilling to undergo surgery

12 Problems with micturition

Linda Cardozo and Cornelius Kelleher

12.1 Introduction

Urinary incontinence is a condition affecting millions of women and placing an immense financial burden on health care resources. Although rarely life threatening, it adversely affects many aspects of the quality of life of sufferers. Incontinence may be secondary to transient and often reversible causes such as urinary tract infection. However, many cases are chronic and last indefinitely unless properly investigated and treated. Appropriate investigation and treatment may not always provide a cure but will improve the quality of life of all sufferers.

12.2 Background

Urinary incontinence is a common and distressing condition. The International Continence Society has defined incontinence as 'a condition of involuntary loss of urine that is a social or hygienic problem and is objectively demonstrable'.

Prevalence studies have estimated that 14% of women in Britain suffer from urinary incontinence, yet such studies are hampered by a reluctance to admit to this disabling condition. An epidemiological study carried out in Greater London, found that 11.5% of women over the age of 65 suffer urinary leakage, but 71% of those suffering moderate to severe leakage were receiving no help from the Health or Social Services. There are many reasons for this reluctance to seek help. The embarrassment and shame suffered by many women as well as the lack of education and information regarding incontinence and its treatment appear to be major factors. Women are often unaware that effective treatment exists and many doctors appear reluctant to discuss urinary problems. A recent national survey has shown that there is a lack of adequate training at undergraduate and postgraduate level in the understanding and management of incontinence.

Urinary incontinence affects all age groups. Despite a persistent myth, incontinence is not a normal part of ageing, although age-related changes of the lower urinary tract predispose to incontinence in the elderly. Urinary incontinence is, however, a common cause of admission to geriatric units and old people's homes because of its association with stroke, cerebrovascular disease and dementia.

Most children become dry between the ages of two and four years although 10% of five year olds are still incontinent, mainly at night (Figure 12.1). Lower urinary tract symptoms are common during all stages of pregnancy and include both irritative bladder symptoms such as urgency, frequency, and nocturia, as well as stress incontinence.

Incontinence affects many aspects of the lifestyle of sufferers. Social, domestic and work lives are often disrupted and sport and leisure activities may be impossible. Otherwise fit and healthy women become house bound and socially incapacitated by the constant need to change pads and carry around spare underwear. Sexual intercourse may be affected by incontinence at penetration or at orgasm and interpersonal

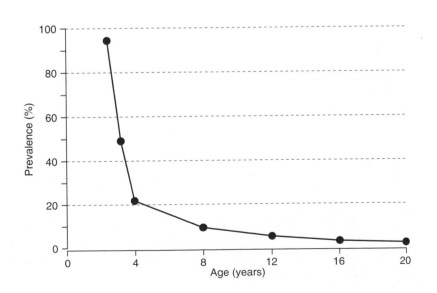

Figure 12.1 The prevalence of nocturnal enuresis by age.

relationships can suffer as a result. Diminished self-esteem and psychological distress are common although the degree of incontinence does not appear to correlate well with the distress it causes. Many women eventually seek help after many years of suffering, and are reluctant to tell even their spouse of their condition.

12.3   The mechanism of continence in women

Under normal circumstances the bladder and urethra function as a reservoir for the storage of urine until voiding becomes necessary and socially acceptable. For continence to be maintained the urethral pressure must exceed the intravesical pressure. Reversal of this balance physiologically during voiding and pathologically during incontinent episodes results in urinary leakage.

12.3.1   Anatomy

The bladder is made of a syncytium of smooth muscle arranged in whorls and spirals called the detrusor, and is adapted for mass contraction. It has been described as a structure of three smooth muscle layers, although frequent interchange of muscle fibres between layers makes them anatomically indistinct. The muscle is lined by a loose and readily distensible mucous membrane, covered by transitional epithelium. The trigone is a triangular area lying between the bladder neck and the ureteric orifices. It is smooth walled and fixed. The smooth muscle of the trigone is easily distinguished into two layers, a deep layer similar in many respects to the detrusor, and a superficial layer which may be important in the prevention of ureterovesical reflux during voiding.

The female urethra is 3–5 cm in length. Proximally it is lined by transitional epithelium continuous with that of the bladder, whereas distally it consists of non-keratinized stratified squamous epithelium and is continuous with introital skin. There appear to be two distinct layers in the urethral muscle coat, an inner layer of smooth muscle fibres, divided into a thick inner longitudinal layer and thin outer circular layer, and an outer layer of striated muscle which comprises the rhabdosphincter urethrae (true urethral sphincter). The proximal part of the urethral striated muscle surrounds the smooth muscle coat circularly, being incomplete in the ventral aspect. The distal part of the striated urethral muscle lies in the urogenital diaphragm.

12.3.2 The continence mechanism

For women to be continent the urethra must seal properly (hermetic closure), the urethral pressure must be higher than bladder pressure, and the lower urinary tract must be under voluntary control. Continence is maintained by many different mechanisms including those integral to, and those outside the urethra.

The main sphincter mechanism is found in the middle of the urethra where the smooth and striated muscle components are best developed and the vascular plexus is most abundant. Continence also depends on the inner urethral wall softness and the epithelial surface is folded to allow better apposition of the mucosa. If this seal is defective higher compressive forces are required to maintain continence. Both these factors are less efficient in the elderly.

Urethral pressure decreases with age and the epithelium of the urethra becomes atrophic after the menopause reducing it's sealing properties. Oestrogen receptors have been demonstrated in the urethra and oestrogen therapy combined with an alpha adrenergic receptor agonist have been shown to be useful in treating postmenopausal stress incontinence.

Below the urethra lies the anterior vaginal wall and its connections to the pelvic floor muscles, which course laterally forming a 'urethral hammock'. The stability of this layer allows the urethra to be compressed and therefore closed during increases in abdominal pressure.

As the bladder fills with urine mechanoreceptors in the bladder wall are activated and action potentials run in the parasympathetic nerves to the spinal cord at the level S2 to S4. In healthy women the micturition reflex is organized in the pontine micturition centre. In patients with complete spinal cord lesions bladder filling is not felt, and their micturition reflex is organized at a spinal level. Loss of central inhibition of the detrusor and coordination of the detrusor–sphincter mechanism, results in a low capacity hyperactive bladder. Voiding difficulties due to a loss of coordination of the detrusor and sphincter (detrusor sphincter dyssynergia) are also often present.

12.4 Causes of urinary incontinence

There are many disorders that affect bladder and urethral function. These can be divided into three major groups:

1. Urinary incontinence
2. Urgency and frequency (irritative symptoms)
3. Voiding difficulties

The major causes of urinary incontinence are:

1. Genuine stress incontinence (GSI)
2. Detrusor instability (DI)
3. Overflow incontinence
4. Fistulae
5. Congenital
6. Urethral diverticulum
7. Temporary (e.g. UTI, faecal impaction, drugs)
8. Functional (e.g. immobility)

Genuine stress incontinence and detrusor instability can co-exist.

12.4.1   Stress incontinence

This the commonest symptom with which women present. Stress incontinence is a symptom, or a sign, but not a diagnosis. Genuine stress incontinence (GSI) is a diagnosis made by urodynamic assessment to confirm the loss of urine per urethram due to increases in intra-abdominal pressure, in the absence of detrusor contractions. It is the commonest cause of urinary incontinence in women accounting for about 50% of cases. Stress incontinence is most commonly caused by coughing, sneezing or, in severe cases, by only minimal activity such as walking. It mainly occurs in parous women and is made worse by lifting, straining and constipation.

The most common mechanism of GSI is thought to be urethral hypermobility due to pelvic floor weakness. Weakness of the pelvic floor allows the proximal urethra to descend below the level at which passive abdominal pressure can be transmitted to it so the additional external pressure on the urethra that facilitates closure is lost. There is histological and electromyographic (EMG) evidence that denervation of the pelvic muscles and damage to the urethra as a result of parturition are important in the aetiology of this condition.

12.4.2   Detrusor instability (DI)

This is the second commonest cause of incontinence. The incidence increases with age and DI is the commonest cause of urinary incontinence in the elderly. The condition is termed detrusor hyperreflexia if there is overt neurological pathology.

An unstable bladder is one which contracts involuntarily. Urodynamic assessment is required for diagnosis. Women usually present with multiple symptoms, most commonly urgency (rapid onset of the desire to void), urge incontinence, frequency and nocturia.

Common causes of frequency and urgency

Gynæcological/urological
Urinary tract infection
Detrusor instability
Bladder lesion (cancer or stones)
Inflammation (e.g. interstitial
   cystitis)
Fibrosis (e.g. after radiation)
Atrophy (menopause)
External pressure (e.g. pelvic
   mass, fibroids)
Prolapse
Pregnancy

Medical/psychological
Drugs (e.g. diuretics)
Diabetes
Neurological conditions
   (e.g. multiple sclerosis)
Excessive fluid intake
Heart failure
Habit

The pathophysiology of detrusor instability is poorly understood, and an underlying cause for the condition is rarely found. In the majority of cases therefore the term idiopathic detrusor instability is used. Detrusor instability and urethral sphincter incompetence (genuine stress incontinence) can occur together, and detrusor instability can arise *de novo* after surgery for stress incontinence.

Sensory urgency is diagnosed when there are intense irritative bladder symptoms in the absence of unstable detrusor contractions. Patients have a reduced functional bladder capacity, daytime frequency, nocturia and sometimes incontinence. It may be caused by inflammatory conditions of the bladder or urethra (e.g. interstitial cystitis, atrophic urethritis), and can be diagnosed by cystoscopy and bladder biopsy.

12.4.3 Overflow incontinence

This usually occurs secondary to chronic retention of urine and is characterized by voiding difficulties, impaired bladder sensation, recurrent urinary tract infections, and the frequent leakage of small amounts of urine.

The causes of overflow incontinence of urine

Acute retention	Secondary to surgery (any operation particularly gynaecological and rectal surgery)
	Secondary to pain (e.g. Herpes genitalis)
Drugs	Tricyclic antidepressants, anticholinergic agents, alpha-adrenergic agonists, epidural analgesia.
Urinary tract infection	
Urethral stricture	Surgery for GSI
	Urethral surgery (e.g. dilatations)
	Vaginal surgery
	Recurrent urinary tract infections
	Radiotherapy
Pelvic mass	Fibroids
	Faecal impaction
Cystocele	
Detrusor hypotonia	Lower motor neurone lesions (e.g. diabetes mellitus)
Psychogenic	Dementia

12.5 How the history can help

Although important, the history and physical examination alone are often inadequate for the accurate diagnosis of lower urinary tract disorders.

When symptoms are limited to those of stress incontinence alone there is a high probability that urodynamic assessment will confirm genuine stress incontinence (GSI). Even the clinical demonstration of stress incontinence, however, cannot exclude the presence of detrusor instability. History and physical examination have a very poor positive predictive value for the diagnosis of other causes of lower urinary tract dysfunction and cannot confidently exclude the presence of voiding difficulties. Although this may be an acceptable margin of error for some patients prior to conservative treatment, it is wholly unacceptable prior to surgical intervention, as the consequences of misdiagnosis and unnecessary or inappropriate surgery are irreparable.

12.6   Why perform urodynamic investigations?

Urodynamic investigations provide an accurate diagnosis of lower urinary tract dysfunction.

Situations where urodynamic investigations are important	
Before surgery	Incontinence surgery
	Symptomatic patients before other gynæcological surgery
	After previous failed surgery
Mixed symptoms	
Failed response to conservative measures	
Voiding difficulties	Previous surgery
	Previous anticholinergic drugs
	Neurological causes
Neurological diseases	
Dementia	
Mentally retarded	
Nocturnal enuresis	
Vesico-ureteric reflux	
Elderly	Non-transient causes
Audit	
Research	

The success of surgery depends on the preoperative assessment, selection and preparation of patients. Not only is it important to diagnose and evaluate stress incontinence but the exclusion of detrusor instability and awareness of voiding difficulties helps prevent major postoperative complications and improve surgical success. The ability to objectively assess surgical outcome by urodynamic investigation is also important for audit and to compare different surgical procedures. This is particularly helpful in view of the number of operative techniques available for the treatment of GSI.

Preoperative urodynamic assessment should not be reserved for those undergoing incontinence procedures alone. If, for instance, urinary symptoms are present in addition to other gynæcological problems for which major pelvic surgery is planned then a preoperative urodynamic assessment is justified to identify any underlying bladder dysfunction. Similarly, urodynamic assessment is essential for the management of patients with neurological conditions where disorders of the lower urinary tract unrelated to but compounding the neurological disability may be present.

12.7   How to investigate bladder function

Investigations can be divided into three categories, simple (suitable for the GP/outpatient setting), basic urodynamic (suitable for the urodynamic unit), or complex urodynamics usually performed at major referral centres.

Simple investigations	Basic urodynamics	Complex urodynamics
Mid-stream urine (MSU) Frequency/volume chart Pad test	Cystometry Uroflowmetry	Videocystourethrography Urethral pressure profilometry Electromyography (EMG) studies Ambulatory urodynamics Ultrasound

Not all tests are required for all patients and a simple guideline to choosing the most appropriate investigations is shown below.

Which test for which patient?	
MSU	All patients
Frequency/volume chart	All patients
Urine flow rate	All patients, esp. voiding difficulties
Ultrasound residual volume	Voiding difficulties
Subtracted cystometry	All patients with non-transient dysfunction
Videocystourethrography	Where available, combined with cystometry
Urethral pressure profile	Voiding difficulties, failed surgery, urethral pathology, neurological voiding dysfunction
Pad test	Adjunct to diagnosis and follow-up of GSI
Ambulatory urodynamics	Not fully assessed

12.7.1   Simple investigations

Urinary tract infection is a common cause of urinary symptoms and therefore the microbiological culture of a mid stream urine sample should be the preliminary investigation for all patients.

(a)   Frequency/volume charts

These provide a simple objective assessment of fluid balance over a period of time. Women are asked to measure their fluid input and output over the course of a week (or less) (Figure 12.2). Recording events over a number of days increases the likelihood of obtaining representative data. Excessive fluid input is often documented and is an easily rectifiable cause of frequency. The volume and frequency of daytime and nocturnal voiding, provide insight into the nature and

Figure 12.2 A urinary diary showing daytime frequency and normal fluid output. W = the timing of incontinence.

Time	Day 1 Input	Day 1 Output		Day 2 Input	Day 2 Output		Day 3 Input	Day 3 Output
0700 hours	250	150		200	160	W	250	170
0800 hours		75	W		50			75
0900 hours	200	140		200	55		150	60
1000 hours		100			70			
1100 hours	150			150		W	200	55
1200 hours		60	W		100			60
1300 hours	100	55		100	50			
1400 hours		75						
1500 hours			W	100				
1600 hours	100							
1700 hours								
1800 hours								

degree of the woman's symptoms and repeated charting provides a simple objective assessment of improvement following treatment.

(b) Pad testing

This is a simple non-invasive quantitative test to assess urinary leakage. A preweighed perineal pad is worn for a period of time, during a number of manoeuvres designed to provoke urinary loss after the bladder has been filled via a urethral catheter to a known volume (usually 250 ml). A number of manoeuvres designed to provoke urinary leakage (e.g. walking, climbing stairs, coughing) are performed after which the pad is reweighed. Incontinence is measured as a significant increase in the weight of the pad. The test can be conducted over various time periods from the standard 1 h ICS (International Continence Society) test to the more prolonged but possibly more reproducible and physiological 24 and 48 h tests.

12.7.2   Basic urodynamic investigations

(a)   Uroflowmetry

This is the measurement of urine flow and provides an objective assessment of voiding ability. The test is rapid, non-invasive, physiological and requires little specialist equipment apart from a commode and a

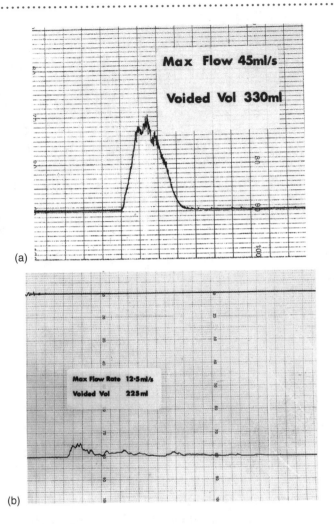

(a)

(b)

Figure 12.3 Uroflowmetry. (a) A normal recording. (b) An abnormal recording showing a prolonged flow time and reduced flow rate.

flow meter. It is indicated when the patient complains of voiding difficulties or prior to surgery. Uroflowmetry allows the measurement of voided volume, maximum flow rate, acceleration of flow rate, average flow rate, flow time, and time to maximum void. A normal and prolonged flow rate are shown in Figure 12.3.

Urinary flow rate varies significantly with the voided volume. A normal flow rate is greater than 15 ml/s provided that at least 150 ml has been voided. There is also significant variability depending on age, sex, strength of detrusor contraction, and previous urethral instrumentation. Nomograms have been established to provide a normal reference range for the maximum and average urine flow rates over a wide range of voided volumes for both men and women.

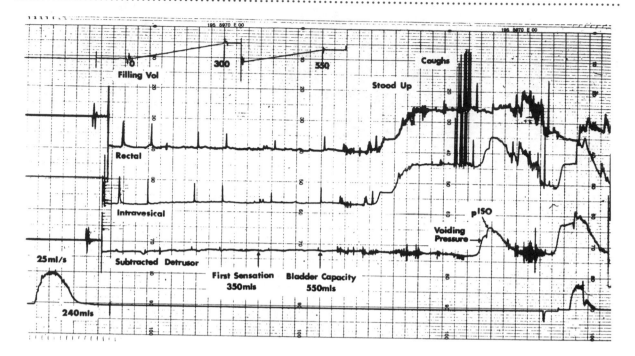

Figure 12.4 A normal cystometry trace.

(b) Cystometry

This is a method by which the pressure/volume relationship of the bladder is measured, and is important to evaluate physiological or pathological detrusor contractions. It is also used to assess bladder sensation, capacity and compliance.

The most commonly used technique for performing cystometry is the retrograde filling of the bladder with fluid via a urethral catheter, measuring the intravesical pressure, and generating a pressure volume curve of the results. A normal cystometry trace is shown in Figure 12.4. Saline is used to fill the bladder, or a radio-opaque contrast medium if radiological imaging of the bladder is to be performed. Either internal or external pressure transducers are used to measure intravesical, and in the case of subtracted cystometry detrusor, pressure. Simultaneous measurement of abdominal pressure by means of a rectal or vaginal transducer allows calculation of the detrusor pressure by subtraction of the intra-abdominal from the intravesical pressure recording. The calculation of detrusor pressure in this manner results in more accurate and reproducible investigations.

The cystometric examination has two phases, the filling phase, followed by the voiding phase.

Filling phase   During filling the normal bladder behaves in a compliant manner with little pressure rise (<15 cmH$_2$O). The patient should be able to describe the first sensation of filling and the first desire to void (>100 ml infused volume), as well as the sensation of fullness at bladder capacity (adult 400 ml).

Causes of poor compliance (a tonic detrusor pressure rise >15 cm H$_2$O during filling) include chronic inflammation or postradiotherapy fibrosis of the bladder. Many cases are idiopathic. If a detrusor contraction can be elicited it is important to see whether the contraction is voluntary. Involuntary contractions are normally absent. When they do occur they are referred to as detrusor instability (or detrusor hyperreflexia if present in conjunction with a neurological abnormality). A cystometry trace showing detrusor instability is shown in Figure 12.5.

Voiding phase   The second phase is the voiding phase which occurs by coordinated activity between the detrusor contraction and sphincter relaxation, the urethral sphincter usually relaxing before the onset of detrusor activity. Voiding is normally to completion with no urine remaining in the bladder after emptying.

The demonstration of a bladder contraction is important because it implies intact innervation of the bladder. A bladder that fails to contract may do so due to a neurological deficit, e.g. diabetes or previous pelvic surgery. Some women can void by urethral relaxation alone, and do not require detrusor contraction.

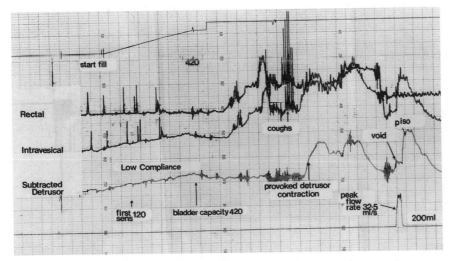

Figure 12.5 A cystometrogram showing systolic detrusor instability.

Basic urodynamics is one of the procedures that you should try to see done
during a clinical attachment.

Cystometry plays the central role in urodynamic assessment. It can, however, be aided by the synchronous use of radiological imaging to visualize the bladder and urethra at the time of disordered function. In a similar way simultaneous EMG studies can be used to evaluate sphincter function.

12.7.3 Complex urodynamic investigations

(a) Videocystourethrography (VCU)

VCU, fluoroscopic examination of the lower urinary tract, in conjunction with intravesical (and intraurethral) pressure measurement is considered to be the 'gold standard' for the diagnosis of lower urinary tract dysfunction and can be recorded on a video-cassette for later reviewing. The purpose of the combined video study is to facilitate viewing of bladder activity while simultaneously monitoring intravesical pressure and sphincter activity. Generally, these techniques are only available in specialized referral centres.

VCU is particularly useful in assessing neurological injury, and in evaluating suspected detrusor sphincter dyssynergia or other types of obstructive uropathy. It is also important to detect anatomical abnormalities such as trabeculation, ureteric reflux (Figure 12.6) or urethral diverticula.

(b) Urethral pressure profilometry (UPP)

The maintenance of continence requires that urethral pressure exceeds intravesical pressure at all times except during micturition. The variation in pressure at different points along the urethra can be recorded by slowly withdrawing a microtransducer tipped catheter from the bladder. A typical urethral pressure profile catheter is shown in Figure 12.7.

A urethral stress profile can be measured using a catheter with two transducers mounted 6 cm apart. One transducer remains in the bladder and the patient is asked to cough repeatedly, while the other is withdrawn along the urethra. The stress profile is merely the subtraction of intravesical pressure from urethral pressure and is an assessment of the net urethral closure pressure during stress manoeuvres. A resting urethral pressure profile can be recorded at maximal urethral pressure to look for urethral instability. A normal urethral pressure profilometry trace is shown in Figure 12.8.

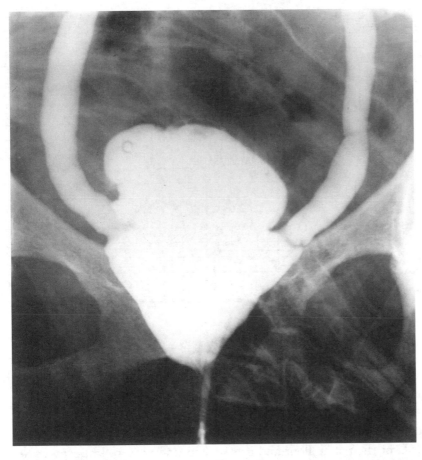

Figure 12.6 Ureteric reflux demonstrated by videocystourethrography.

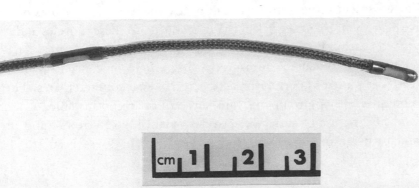

Figure 12.7 A urethral pressure profilometry catheter showing two microtip transducers 6 cm apart.

Unfortunately despite appearing to be an ideal investigation, studies of urethral pressure measurement have shown that the overlap between normal and GSI patients is too large to allow the test to be used to diagnose incontinence. Urethral pressure profilometry has, however,

Figure 12.8 A normal urethral pressure profilometry trace.

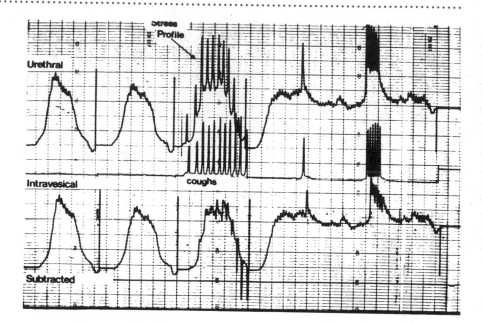

other applications, namely the assessment of voiding difficulties, particularly those due to outflow obstruction (e.g. strictures), and the diagnosis of a urethral diverticulum.

(c)   Electromyography

EMG of the pelvic floor and urethral sphincter is an important part of the assessment of neurological voiding dysfunction. It provides objective data about the integrity of the innervation to these muscles and the synchronization between detrusor and external sphincter.

EMG activity can be measured using single or concentric needle electrodes placed in periurethral, or external anal sphincter sites. These techniques although superior are sometimes uncomfortable. Understandably the use of surface electrodes, anal plug electrodes, and urethral catheter mounted electrodes despite being less accurate, has become popular as they are painless, and easy to apply.

(d)   Ultrasound scanning (USS) and magnetic resonance imaging (MRI)

These techniques have been advocated as non-invasive, inexpensive alternatives to videocystourethrography. Transabdominal scanning is limited in value due to the position of the bladder neck behind the symphysis pubis, but is a useful test to determine the residual urine volume following voiding and to investigate morphological abnormalities.

Vaginal scanning is unacceptable to some patients and has the added disadvantage of restricting urethral mobility and thus distorting anatomy. Perineal scanning has the obvious benefit of being a less invasive and probably more acceptable procedure whilst providing excellent imaging of the bladder and urethra.

MRI imaging of the lower urinary tract has so far been limited to a research setting. The advantage over USS lies in its ability to provide good soft tissue differentiation. The test is expensive, and at present MRI imaging cannot be considered a practical aid to the diagnosis of female urinary incontinence.

(e) Ambulatory urodynamics

Due to the unphysiological nature of conventional urodynamic investigations, and the inability to consistently reproduce urinary symptoms during the investigation the idea of long-term ambulatory urodynamics is attractive. The test still requires an indwelling urethral or suprapubic catheter to record intravesical pressure. The long-term monitoring facility allows natural bladder filling to take place and also enables patients to take part in the everyday activities which may precipitate their urinary symptoms.

The technique, although needing further evaluation, has been shown to be a sensitive index of lower urinary tract dysfunction in patients with symptoms of detrusor instability but negative cystometry. Unfortunately 'normal' individuals may also have detrusor contractions on ambulatory monitoring, and thus interpretation of this finding must be approached with caution.

12.8   Management of urinary incontinence

In general practice, women will be managed conservatively by a combination of physical and behavioural techniques, or with the use of specific drugs. For some patients the provision of incontinence aids will be necessary, and this is often coordinated by a district continence advisor or suitably trained district nurse. Outlines of the management for genuine stress incontinence and detrusor instability are shown below.

Management of genuine stress incontinence

Conservative	Surgical
Pelvic floor exercises (with or without biofeedback)	Vaginal anterior colporrhaphy
Weighted vaginal cones	Abdominal bladder neck suspensions, i.e. Marshall Marchetti, Burch, Aldridge
Pelvic floor 'functional electrical stimulation'	Endoscopic bladder neck suspension, i.e. Stamey
Oestrogen therapy	Periurethral collagen injections
Alpha adrenergic drugs, e.g. phenylpropanolamine	Artificial sphincter

12.8.1 Conservative management of genuine stress incontinence (GSI)

For many patients conservative measures will be the best option. Women with only mild GSI will often respond well to conservative treatment alone and those unfit for surgery are often significantly improved. It is important to ensure that patients have completed their families prior to embarking on surgical treatment as primary surgery is usually the most successful, and vaginal delivery often results in a return of incontinence for those who have embarked on an early surgical cure.

(a) Pelvic floor exercises (PFEs)

These are still the mainstay of physiotherapy treatment for GSI. PFEs are often performed incorrectly and hence women must receive appropriate instruction on how to do them. A perineometer can be used to give the patient biofeedback regarding her ability to perform these exercises (Figure 12.9). This instrument consists of a vaginal resistance chamber attached to a meter which registers an increase in pressure when the appropriate muscles contract. Alternatively women can be taught to place their fingers in the vagina to ensure adequate contraction of the levator muscles. Various authors have confirmed the merits of these exercises in patients with mild to moderate GSI.

(b) Vaginal cones

These are a set of cone-shaped weights and were first introduced in 1985 (Figure 12.10). It was shown that women could be trained to

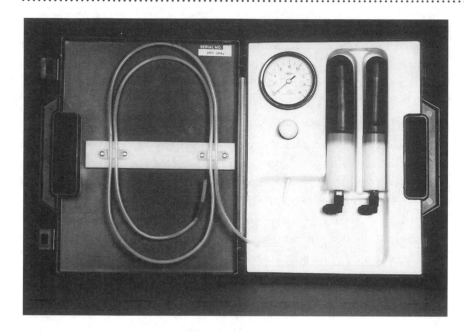

Figure 12.9 A Bourne perineometer.

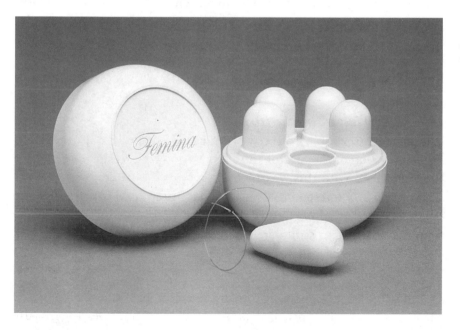

Figure 12.10 A set of vaginal cones. (Note the thread attached to the cone to facilitate removal from the vagina.)

contract and strengthen their pelvic floor muscles in order to retain cones of increasing weight in the vagina. Patients are instructed to retain the cone in the vagina for 15 min twice a day, successively increasing the weight of the cone. By strengthening the pelvic floor in this manner many women find their symptoms cured or significantly improved. Women may find the cones easier to use than performing pelvic floor exercises although they are not useful when women have a significant degree of vaginal prolapse.

(c)   Other devices

Various different devices have been used over the years, although many are now largely of historic interest only. Devices are designed to compress or occlude the urethra, and are intended primarily for women with stress incontinence which is exacerbated by, or associated with only certain activities, but not problematic for the majority of the time.

A vaginal sponge, or tampon, compresses and elevates the urethra and offers some symptomatic relief, although repeated or prolonged usage can result in vaginal discomfort due to the absorbent nature of these products.

Perhaps the most promising devices to have been investigated recently are the urethral plugs. These silicone urethral inserts have a short string attached to facilitate removal and a small inflatable balloon to keep the device in position. Although not an attractive option to all women, trials in the United States of America have shown satisfaction with the technique by many women with minimal side-effects.

(d)   Electrical stimulation

This appears to work by directly stimulating the pudendal nerve causing contraction of the levator muscles. Various degrees of success have been described with this treatment.

Faradism and interferential therapy have been used to stimulate the pelvic floor as an adjunct to pelvic floor exercises. Both utilize a low frequency current to stimulate the levator ani muscles.

(e)   Oestrogen replacement therapy

Oestrogens play an important role in the maintenance of continence by their effects on the functional layers of the urethra. Oestrogens also appear to potentiate the effect of alpha adrenergic receptor agonists (e.g. phenylpropanolamine) on the urethral closure pressure. There is no significant evidence that oestrogen therapy alone significantly improves stress incontinence, although combined with other treatments it is of benefit for those awaiting or unsuitable for surgery.

12.8.2   Surgical treatment of genuine stress incontinence (GSI)

Women who remain symptomatic despite conservative therapy or who have a major degree of GSI will require surgical treatment.

Well over a hundred different operations for the treatment of GSI have been described and performed with varying degrees of success. The most common surgical procedures are as follows:

The surgical management of genuine stress incontinence

Operation	Indication
Colposuspension	Primary or secondary GSI with cystocele
Marshall Marchetti Krantz	Primary or secondary GSI
Sling procedures	Recurrent GSI
Long needle suspensions (e.g. Stamey)	GSI in the surgically difficult pelvis
Anterior repair	Severe anterior vaginal wall prolapse and GSI
Periurethral injectables (e.g. collagen)	Recurrent GSI or medically unfit women
Complex surgical procedures (e.g. neourethra, artificial sphincter)	Intractable recurrent GSI

(a)   Colposuspension

Retropubic dissection is required to mobilize the bladder neck medially off the underlying fascia, prior to the insertion of two to four long-term absorbable or non-absorbable sutures from the paravaginal fascia to the ipsilateral ileopectineal ligament. The complications of this procedure include operative blood loss, urinary tract damage and infection, and the later problems of voiding difficulties, detrusor instability and enterocele formation. The long-term cure rate for colposuspension, however, is superior to that of other operations and this procedure remains the one of choice for treatment of GSI.

Recently in order to reduce patient morbidity, and convalescence, laparoscopic colposuspension has been developed utilizing either an extraperitoneal or transperitoneal approach to the retropubic space. As yet despite evidence of reduced short-term patient morbidity, no long-term follow-up data support the efficacy of this approach. It is possible that early mobilization following surgery may even decrease the efficacy of colposuspension although this too has yet to be demonstrated.

(b)   Marshall–Marchetti–Krantz (MMK)

This is the predecessor of the colposuspension. A non-absorbable suture material is used to take a double bite of tissue from the bladder neck and hitch it up to the back of the pubic bone. Although this is a very successful operation for the treatment of GSI, it does not treat anterior vaginal wall prolapse and osteitis pubis may be a postoperative problem occurring in about 5% of cases.

(c)   Colposuspension or Marshall–Marchetti–Krantz procedures

Primary or recurrent GSI can be treated using these procedures performed through a transverse suprapubic abdominal incision. They can be used when there is mobility of the bladder neck allowing sufficient elevation and repositioning to be achieved by the suspension procedure. Anterior vaginal wall prolapse can also be corrected by a Burch type of colposuspension.

(d)   Anterior colporrhaphy (anterior repair)

This has been used to treat primary genuine stress incontinence in the presence of a cystourethrocele, but although it is the best operation for the treatment of anterior vaginal wall prolapse it is not the best treatment of GSI.

The technique involves an anterior longitudinal vaginal wall incision, mobilization of the bladder neck and the insertion of supporting sutures to elevate the bladder neck, prior to closure of the pubovesical fascia and the anterior vaginal wall. The operation has few complications, is quick and easy to perform and requires a short hospital stay, but the long-term cure rate of GSI is poor.

(e)   Endoscopically guided bladder neck suspension

Many different endoscopic procedures have been described, but all utilize two nylon sutures passed blind from the paravaginal fascia to the rectus sheath or vice versa. They are easier to perform than conventional surgery in the surgically difficult pelvis where recurrent surgery has made the usual anatomical landmarks less discernible. They can be used for primary or recurrent GSI and have been advocated for the treatment of the elderly or infirm, as they are quick and easy to perform, do not always require a general anaesthetic, and require only a short hospital stay. Long-term cure rate is less than that following colposuspension or MMK.

(f) Slings

These procedures are advocated for the treatment of recurrent GSI especially where there is limited mobility of the anterior vaginal wall and bladder neck, and a deficient sphincter, often as a result of previous surgery. Many different types of sling operations have been described. The sling can be fashioned from organic material (bovine rectus sheath), strips of the patient's own rectus sheath, or inorganic material such as Mersilene or Gore-tex. Sling insertion is via an abdominal incision with retropubic dissection, or through a vaginal incision but often a combination of the two is used. Cure rates appear to be high, although complications including voiding difficulties are common following these operations.

(g) Injectables

The injection of substances alongside the urethra and bladder neck has been introduced for the management of women with a fixed scarred urethra following previous surgery, or for those who would benefit from a short and simple operative procedure. Various injectables have been used, and although initial attempts with teflon were hampered by migration of the teflon to distant sites, results with glutaraldehyde crosslinked bovine collagen (GAX) and microparticulate silicon appear promising and problems with migration have not been encountered. These techniques have a high improvement rate but low cure, are expensive and await more thorough evaluation.

(h) Complex procedures

For women who have failed all attempts to correct their incontinence several complex procedures exist. A neourethra can be fashioned from a flap of bladder or an artificial urinary sphincter implanted. These operations are complicated, and when all else has failed a urinary diversion may be the final answer for a small minority of patients.

12.8.3 Choice of surgery

The choice of surgery will often be dictated by the experience and preference of the operator, previous surgery, the presence of prolapse, and the clinical condition of the patient. For the fit patient with primary or recurrent GSI colposuspension will be the preferred procedure for the majority of operators.

12.9 Treatment of detrusor instability (DI)

Treatment of DI aims either to improve central control of the voiding reflex, or to alter the peripheral innervation of the detrusor and prevent detrusor contractions.

Treatment of detrusor instability

Behavioural intervention
Bladder training (bladder drill)
Biofeedback
Hypnotherapy and acupuncture
Maximal electrical stimulation

Drugs

Name	Effect
Oxybutinin	Anticholinergic and direct smooth muscle relaxant. 2.5–10 mg bd or tds
Propantheline	Anticholinergic. 15–90 mg qds
Imipramine	Anticholinergic, muscle relaxant, possible alpha receptor agonist. 50 mg bd up to 150 mg as a single dose
DDAVP	Reduces urine production. 20–40 µg nocte as a nasal spray.
Oestrogens	Raise the sensory threshold of the bladder

Surgery
Augmentation cystoplasty (CLAM)
Urinary diversion

Bladder retraining (bladder drill), is a form of behavioural therapy that aims to alter a pattern of maladaptive learned behaviour. It may be performed as an inpatient or outpatient procedure; the former is more costly but also more effective. Strict timed voiding is introduced, starting with an interval of 1 or 1.5 h. Over a period of days or weeks the voiding interval is increased and a normal voiding pattern can be achieved. Patients need encouragement, and sometimes reduction of fluid intake.

Both hypnotherapy and acupuncture have been shown to be effective in the treatment of DI but are time consuming and therefore costly, and studies so far have been uncontrolled. Biofeedback uses a visual or tactile signal to increase awareness of detrusor activity, and allows the patient to inhibit detrusor contractions.

Maximal electrical stimulation of pudendal nerve afferents via anal or vaginal plug electrodes results in inhibition of detrusor contractions. Portable, but expensive, stimulators have been developed, and varying degrees of success have been described with this technique.

The majority of patients will be managed with drugs either alone or combined with other forms of treatment. Most of the drugs have anticholinergic properties and result in relaxation of the detrusor musculature. Anticholinergic side-effects are often unacceptable and restrict treatment compliance in some patients. The most commonly prescribed drugs are oxybutynin (Ditropan/Cystrin), propantheline (Probanthine), or imipramine (Tofranil). New drugs which offer greater specificity for the M3 muscarinic receptors, and therefore fewer systemic side-effects, are currently being evaluated in clinical trials.

DDAVP (1-deamino, 8-arginine vasopressin) is effective in women with intractable nocturia, or nocturnal enuresis. It is administered nasally at bed-time and is known to reduce nocturnal urine production by up to 50%. Caution must be exercised in the use of this drug in the elderly, where fluid retention and cardiac compromise can be a serious problem.

Oestrogens may alleviate the symptoms of frequency, urgency, urge incontinence, dysuria, and nocturia. They also improve the quality of life of many peri- and postmenopausal women.

Conventional bladder neck surgery is not helpful in the management of detrusor instability. Vaginal denervation, bladder transection, sacral neurectomy, cystodistension, and phenol injections have all been used with limited success, but have largely been replaced due to unacceptable side-effects.

For the small minority of patients with severe DI uncontrolled by conservative means, augmentation cystoplasty (CLAM) remains the surgical procedure of choice. This involves the augmentation of the bladder using a patch of bowel. The procedure requires major surgery and is often complicated by voiding difficulties requiring long-term intermittent catheterization. The long-term risks of neoplasia in the bowel segment will also need to be evaluated.

A special group of patients are those with both detrusor instability and urethral sphincter incompetence (USI). Primary bladder neck surgery invariably causes deterioration of their irritative bladder symptoms, and the initial aims of treatment are, therefore, to control the detrusor instability. Following successful management, reassessment and if necessary surgery for their urethral sphincter incompetence can be performed.

12.10   The role of continence advisors

For some women the provision of incontinence aids will be necessary. Absorbent incontinence products are not prescribable and are therefore obtained through the community nursing services. The cheapest pads are usually less effective and often prove to be a more costly long-term option. Pads are usually worn in close fitting pants, but in cases of mild incontinence the patient's own underwear is effective and often more comfortable.

Some elderly or disabled patients will be managed in the community by catheterization, either with an indwelling urethral or suprapubic catheter, or by means of clean intermittent self catheterization (CISC). The most common indications are acute and chronic retention, postoperative bladder drainage, the neuropathic bladder and intractable urinary incontinence.

Whichever method is adopted, close supervision by a suitably trained district nurse, or specialist continence advisor is required to avoid complications.

Learning points

Urinary incontinence can be a devastating condition which adversely affects many aspects of the quality of life of sufferers.

An understanding of the continence mechanisms and how various conditions may affect them is an essential prerequisite to proper evaluation and management.

Appropriate investigation and treatment can alleviate the majority of suffering for these women and allow them to return to a normal lifestyle.

Urodynamic investigations are vitally important in accurate assessment, particularly in cases of detrusor instability.

Both conservative and surgical management methods are important.

Elevation of the bladder neck is the basic surgical principle in the treatment of genuine stress incontinence.

Behavioural therapy and anticholinergic drugs are the basics of managing detrusor instability.

Detrusor instability and genuine stress incontinence can co-exist.

Further reading

Cardozo, L.D. and Cutner, A. (1992) Surgical management of genuine stress incontinence. *Contemp. Rev. Obstet. Gynaecol.*, 4, 36–41.

Gilpin, S.A., Gosling, J.A., Smith, A.R.B. and Warrell, D.W. (1989) The pathogenesis of genitourinary prolapse and stress incontinence of urine. A histological and histochemical study. *Br. J. Obstet. Gynaecol.*, 96, 15–23.

Health Survey Questionnaire 1990. Market and Opinion Research International (MORI), 95 Southwark Street, LONDON SE1 0HX 3.

Thomas, T.M. *et al.* (1980) Prevalence of urinary incontinence. *Br. Med. J.*, 281, 1243–5.

13 Menopausal problems

Elizabeth Payne

13.1 Definitions

The **menopause** is the last menstrual period. In the United Kingdom this occurs at the mean age of 51–52 years.

The **climacteric** is the period of progressive ovarian failure which precedes the menopause and usually lasts about five years. During this time the primordial follicles become increasingly resistant to the stimulation of follicle stimulating hormone (FSH), the production of oestrogen declines and the pituitary production of FSH rises due to the negative feedback mechanism.

The climacteric and the postmenopause are relative **oestrogen deficiency** states.

The life expectancy of a woman reaching the menopause is 81 years, therefore a woman can expect to spend a third of her life in the postmenopause.

13.2 Symptoms of the menopause

13.2.1 Short-term symptoms

These are a consequence of the rapidly falling oestrogen levels and the incidence is related to the rate of decrease. With a surgical menopause 100% of women will experience symptoms but with the natural menopause about 82% of women experience acute symptoms for more than one year and in 25% of women the symptoms last for more than five years.

Of those women who experience symptoms, 51% are severe, interfering with daily living, 33% moderately severe and in 16% they are mild.

Short term or climacteric symptoms

Vasomotor symptoms	Hot flushes, night sweats
Menstrual disturbances	Irregular periods, prolonged, heavy periods
Psychological symptoms	Depression, loss of confidence, loss of libido, tiredness

---

Hot flushes and night sweats are the commonest of the climacteric symptoms and in 70% of women persist for more than two years. For many women they prove incapacitating, flushes occurring up to 40–50 times per day, often precipitated by stressful situations and contributing to the undermining of the woman's self-confidence. Similarly the woman may be woken from sleep 7–10 times per night with drenching sweats that necessitate the change of both night wear and bed clothes. As a consequence, both her sleep and that of her partner is regularly disturbed, leading to progressive tiredness, irritability and loss of libido.

Climacteric symptoms are self-limiting and gradually subside as the levels of circulating oestrogens stop fluctuating and finally fall to a low, but steady state.

13.2.2   Medium-term symptoms

Following the menopause, atrophic symptoms begin to occur in many tissues including epithelium and connective tissues.

(a)   Vagina and vulva

Quite soon after the menopause many women notice troublesome vaginal dryness and failure of lubrication. Over the next five to ten years there are progressive atrophic changes of the lower genital tract resulting in loss of vulval architecture, thinning of the vaginal epithelium with an increased susceptibility to infections such as *Candida* and, in extreme cases, vaginal bleeding.

(b) Lower urinary tract

The urethra and bladder trigone also appear to be oestrogen dependent. With the menopause many women experience a wide range of symptoms which include frequency, dysuria, urgency and urge incontinence due to the atrophic changes of the epithelium of the urethra and trigone, an increase in the incidence of urinary tract infections, stress incontinence of urine due to loss of collagen from the proximal urethra and loss of smooth muscle from the urethral sphincter and voiding difficulties which are often associated with the development of a cystocele. In addition there is a loss of tone in the muscles of the pelvic floor and in the older woman there may be prolapse of the urethral mucosa at the meatus (Chapter 12).

(c) Other tissues

The postmenopause is characterized by a generalized progressive loss of connective tissue. In the skin this causes obvious thinning with an increased tendency to spontaneous bruising. In the ligaments it results in non-specific joint pains. There is also a gradual but generalized loss of both muscle mass and muscle strength, a reduction in blood flow to all tissues and a slowing in transmission of nerve impulses. The consequences will vary with the part of the body and the tissue involved but may contribute to loss of vaginal sensation, less frequent orgasm and loss of sexual satisfaction.

Medium term symptoms	
Genital tract	Vaginal dryness
	Vulval and vaginal atrophy
	Susceptibility to infection
	Vaginal bleeding
	Dyspareunia and loss of libido
Urinary tract	Frequency, dysuria, nocturia
	Urgency and urge incontinence
	Stress incontinence
	Prolapse of urethral mucosa
	Urethrocele, cystocele
Connective tissue	Uterovaginal prolapse
	Thinning of skin
	Non-specific muscular and joint pain

13.2.3 Long-term symptoms

These arise after a prolonged period of oestrogen deficiency, and usually become apparent about 10–15 years after ovarian failure. The most important long-term symptoms are cardiovascular disease and osteoporosis.

(a) Cardiovascular disease

Cardiovascular disease and particularly ischaemic heart disease is the leading cause of death in women. Oestrogen appears to confer a protective effect against cardiovascular disease and myocardial infarction (MI) and ischaemic cerebrovascular accidents are relatively uncommon events in the premenopausal woman. Following the menopause and the loss of the protective effect of oestrogen, the incidence of both these conditions increases rapidly, reaching that of men by the seventh decade. For postmenopausal women having a myocardial infarct, the mortality is almost double that for a man of the same age. A woman is twice as likely to die within 60 days of an infarct, more likely to suffer a second infarct with 45% of women dying within a year of an MI.

Oestrogen appears to offer a cardioprotective effect which is independent of age such that a postmenopausal woman of any age is at greater risk than her premenopausal sister of the same age.

The cardioprotective effect of oestrogen is mediated by a tendency to reduce low density lipoprotein (LDL) cholesterol and increase high density lipoprotein (HDL) cholesterol, by a direct effect on the arterial wall causing vasodilation and indirectly by causing a fall in blood pressure.

(b) Osteoporosis

Osteoporosis is characterized by a reduction in the amount of connective tissue and loss of calcium, resulting in reduced bone strength and an abnormal susceptibility to fracture. In the postmenopause, oestrogen deficiency leads to a relative increase in the activity of osteoclasts and reduction in the bone repairing activity of the osteoblasts with resultant bone loss. The effect is most evident in trabecular bone and is most rapid during the first two years after the menopause.

Osteoporosis is a major cause of both morbidity and mortality in postmenopausal women. The most common sites for osteoporotic fractures are:

1. Crush fracture of the vertebrae (1:4 women by age 65 years)
2. Colles' fracture of the wrist
3. Fracture of neck of femur (1:8 women by age 85 years).

The incidence of crush fracture of the vertebra and Colles' fracture of the wrist increase soon after the menopause whereas hip fractures are not common until 10–15 years after the menopause. It is estimated that at least 50% of women will have at least one osteoporotic fracture during her life although this is probably an underestimate as many crush fractures of the vertebrae go unnoticed.

A single vertebral crush fracture usually causes few symptoms and may be recognized only as an incidental finding on radiograph. However, multiple crush fractures will lead to the development of an obvious Dowager's hump with progressively impaired mobility, limitation in inspiratory function and possible predisposition to chest infections.

Colles' fracture of the wrist can usually be treated as an outpatient but, at least in the short term, will significantly impair the ability of the woman to care for herself and may result in permanent disability.

A hip fracture by contrast always necessitates hospital admission which may often be prolonged and may result in permanent loss of mobility such that she may never again lead an independent life. Not only does this have serious implications for NHS resources but also for social services and other agencies in the provision of district nurses, bath attendants, 'Meals on Wheels', and the provision of sheltered accommodation.

It must also be remembered that a hip fracture is a potentially life-threatening condition. About 28% of women who sustain a hip fracture die as a direct result.

13.3   The woman's view of the menopause

The menopause, as a landmark in a woman's life, has been recognized throughout history. In the Bible the association with loss of fertility is described in relationship both to Sarah, wife of Abraham (Genesis 17), and Elizabeth, the mother of John the Baptist (Luke 1).

The menopause is viewed by different women in different ways; these views being influenced by the culture of the society, a woman's personal experience and expectations and premorbid personality. For some it is viewed as a very positive experience, signifying the cessation of possibly troublesome menstrual problems, relief from the constant fear of unwanted pregnancy and elevation to the status of respected elder in society with all the associated privileges. For others, however, particularly in 20th century Western society which places such emphasis on youth and beauty, the menopause can be viewed totally negatively with fear for loss of femininity, loss of attractiveness to the opposite sex and loss of fertility. In these women there may be an associated loss of both self-esteem and self-confidence resulting in depression.

A woman's view of the menopause will also be modified by her personal experience of associated symptoms which, if severe, can have a devastating effect on both the life of the individual woman, her partner and her family and may place a tremendous strain on the marital relationship.

13.4   Approach to management of the menopause

The approach to the management of the menopause has undergone major reform over recent years. Twenty years ago climacteric symptoms were considered by the majority of general practitioners to be a normal part of a 'woman's lot' which she was encouraged 'to put up with' with little support and even less medical therapy. Today most clinicians recognize that climacteric symptoms can have a major impact on the quality of a woman's life and adopt a sympathetic approach to treatment. A smaller number of enlightened doctors will actually offer both counselling and prophylactic treatment for the prevention of long-term sequelae.

13.5   Hormone replacement therapy

Hormone replacement therapy (HRT) is given to combat the effects of loss of endogenous oestrogen activity as a consequence of ovarian failure. Oestrogen may be given alone to women who have undergone hysterectomy but must be given in combination with a progestogen to women who have a uterus. The dosages of oestrogen given are low and are designed to give serum levels comparable to those found in the early follicular phase of the cycle. Because the oestrogen levels are low they do not reliably suppress the pituitary production of FSH and thus do not reliably prevent ovulation.

13.5.1   History of hormone replacement therapy

When HRT was first introduced, in the light of the then current understanding that the climacteric and postmenopause were oestrogen-deficiency states, continuous oestrogen therapy was given alone (unopposed oestrogen therapy). This proved very effective in relieving the symptoms but was associated with an increase in the incidence of both endometrial hyperplasia (7–15%) and endometrial carcinoma. The risk of development of endometrial hyperplasia was subsequently found to be reduced by the addition of a course of a progestogen. This modification also reduced the incidence of breakthrough bleeding and ensured regular endometrial shedding. More recently it has been recognized that it is the length of the course of progestogen and not the total dose that is important in reducing the risk of developing endometrial hyperplasia and it is for this reason that most modern formulations contain a 12-day course of progestogen every 28 days (continuous opposed regime.)

13.5.2   Types of hormone replacement therapy

Types of Hormone Replacement Therapy		
Oral	Daily	Either unopposed oestrogen or combined with a progestogen
Transdermal	Twice weekly	Either unopposed or combined with progestogen patch
Percutaneous	Daily	Unopposed or combined with progestogen
Implant	3–6 monthly	Oestrogen with or without testosterone. Oral progestogens necessary if uterus is *in situ*
Vaginal creams, vaginal controlled-release tablets, rings	Daily or alternate days	Oestrogen may be absorbed systemically and can cause endometrial stimulation

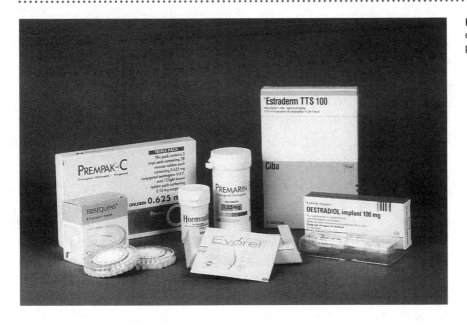

Figure 13.1 A selection of commercially available HRT preparations.

A selection of commercially available HRT preparations is shown in Figure 13.1.

(a) Oral HRT

Oral HRT is simple to administer, cheap and the most widely used form of HRT prescribed today. Different formulations contain different types of 'natural' oestrogens (conjugated equine oestrogens, 17ß-oestradiol, oestriol and oestrone) and different progestogens (levonorgestrol, norethisterone, medroxyprogesterone acetate). The main disadvantage of oral HRT relates to the small incidence of gastrointestinal side-effects and the fact that there is a wide variation in plasma levels achieved due to variations in absorption and hepatic metabolism. The oestrogen is presented as a bolus to the liver. There is theoretically the risk that this will adversely affect the hepatic synthesis of fibrinolytic and clotting factors although in practice this is rarely a problem.

(b) Transdermal

An alternative approach to administration is the use of patches impregnated either with oestrogen or progestogen. The patches are applied to a clean, dry area of skin, usually the lower back or the thighs, twice per week. The steroid from the patch then diffuses directly into the systemic circulation, by-passing the gastrointestinal system and avoiding the hepatic 'first-pass' effect.

Absorption from transdermal patches is reliable, the major disadvantages being that in hot weather or in patients who perspire excessively there is a tendency for the patch to fall off and, with continued use, a small number of patients develop local skin reactions.

(c)   Percutaneous

Oestrogen may be absorbed directly through the skin. A metered dose of oestrogen-containing gel is rubbed into each shoulder and arm. It is absorbed into the stratum corneum within about five minutes. The skin then acts as a reservoir delivering predictable levels of oestrogen directly to the systemic circulation during the course of the day. It is non-greasy, odourless and not visible. It can be used by women who are unable to use patches because of adverse skin reactions and/or poor adhesion.

(d)   Implants

These are small pellets of oestradiol which are inserted into the subcutaneous fat of the anterior abdominal wall under local anaesthetic. Each implant usually lasts 4–8 months.

The advantage of the use of an implant is most apparent in the woman who has undergone hysterectomy, which, once inserted, will free her from the need to take any tablets. This advantage is lost in the woman who retains her uterus who must necessarily be given a 12 day course of progestogen each month.

In certain patients treated with oestradiol implants, climacteric symptoms recur when the oestradiol levels are still well above the postmenopausal range. If oestradiol levels are not monitored, the clinician may repeat the implant each time the woman becomes symptomatic such that the serum levels may increase to almost unrecordable levels. This effect, known as 'tachyphylaxis' due to down-regulation of oestrogen receptors, may result in women complaining bitterly of climacteric symptoms with oestrogen levels comparable to that of pregnancy. In certain cases it has taken more than two years for the levels to fall again into the normal range! This effect can be reduced by limiting the dose of individual implants and avoiding too frequent administration.

(e) *Vaginal creams and vaginal controlled-release tablets*

The vaginal route is usually reserved for those women with genitourinary symptoms. Treatment is usually limited to a short course which is discontinued once the symptoms have resolved and repeated as necessary if symptoms recur. Although theoretically there is a risk of systemic absorption and endometrial stimulation this does not appear to be a problem in practice.

The disadvantage of creams is that they tend to be messy to use with significant leakage from the vagina with soiling of under clothes. This problem has been resolved by the introduction of the controlled-release vaginal tablet.

(f) *Tibolone*

Introduced in 1991, this is a synthetic oral steroid with oestrogenic, progestogenic and androgenic activity. The stimulatory oestrogenic effect on the endometrium is opposed by the progestogenic effect with the result that the endometrium is rendered atrophic. The advantage of tibolone for HRT is that it can be given to a woman with a uterus without the need for a withdrawal bleed.

Use to date suggests that it is tolerated well but because of the androgenic effects it tends to cause greasy skin and unacceptable weight gain.

13.5.3   Choice of route of administration

The choice of the route of administration will depend on the following factors:

1. Patient preference
2. Presence or absence of uterus
3. Need for systemic or local therapy
4. Ability of patient to tolerate withdrawal bleeds
5. Medical factors such as need to avoid 'first pass' effect on hepatic synthesis in women with previous thromboembolic disease, poor oral absorption, skin reaction associated with use of transdermal therapy.

13.5.4   Benefits of hormone replacement therapy

Hormone replacement therapy, either oestrogen alone or in combination with a progestogen is very effective in the relief of all short-term (climacteric) and medium-term symptoms of the menopause provided that an adequate level of replacement is achieved. HRT has also been shown to be very effective in the treatment and prevention of the long-term sequelae of the menopause. Oestrogen replacement therapy can reduce the rate of bone loss and may be capable of increasing bone density, even in established osteoporosis. The use of HRT is associated with a 50% reduction in the incidence of osteoporotic hip fracture and associated mortality.

The use of HRT is also associated with a 60% reduction in mortality from ischaemic heart disease and stroke and even if initiated after the first infarct, appears to be able to reduce the risk of a second.

Contrary to popular belief, HRT is associated with a small fall in blood pressure (probably secondary to its vasodilatory effect) and when compared to non-users of HRT, users of combined HRT have a lower incidence of endometrial carcinoma.

13.5.5   Risks of hormone replacement therapy

Hormone replacement therapy has been widely used in the UK for almost 20 years. Although epidemiological studies suggest that the mortality for users for almost all causes is lower than that for non-users, one question remains unanswered: 'Is there an increased risk of breast cancer in association with the use of HRT?' Several large population studies have recently been published which suggest that there may be a small increase in the risk of developing breast cancer with long-term use. The relative risk is 1.4–1.7 times that of the untreated population after 10–15 years of use.

13.5.6   Duration of therapy

The ideal duration of therapy will depend on the reason for which it was initially prescribed.

Climacteric symptoms, particularly vasomotor symptoms persist for 12 months to more than five years. It is suggested therefore, that treatment is continued for a minimum of two years following which it may gradually be withdrawn provided that symptoms do not recur. If symptoms should recur, treatment should be continued with further attempts at withdrawal every 12–18 months until it can be withdrawn without symptom recurrence.

Whenever HRT is withdrawn it should be done gradually, otherwise climacteric symptoms will be precipitated.

If therapy was initiated with a view to long-term prophylaxis it must be remembered that long-term sequelae take about 10–15 years to become apparent after the menopause. If we assume that on reaching the menopause the woman can expect to live until the age of 80 years, it would seem logical to offer therapy for at least ten years. If the risks of osteoporosis are very high it may be appropriate to continue therapy beyond this time but the benefits must be weighed against the possible increased risk of breast cancer.

13.5.7 Management of common side-effects of HRT

The side-effects of HRT may be due either to the oestrogen or to the progestogen component and are usually dose dependent.

(a) Failure to control menopausal symptoms

If hot flushes or sweats or other menopausal symptoms persist, this is almost always due to insufficient dose or poor absorption of oestrogen. In the first instance the dose should be increased, the ideal dose of oestrogen being the lowest which satisfactorily relieves the symptoms. If this proves ineffective it may be appropriate to change to a different route of administration.

(b) Heavy or prolonged withdrawal bleeds

These are the commonest single cause for non-compliance and discontinuation of HRT. In order to facilitate compliance, all women who have an intact uterus should be fully counselled such that they understand that with standard treatment regimes they will have a monthly withdrawal bleed. This may be heavy initially but usually becomes lighter with time.

Heavy bleeding may respond either to a reduction in the dose of oestrogen or an increase in either the dose or duration of progestogen. Some women are unable to tolerate any vaginal bleeding. Provided that they are at least two years post-menopausal they may be treated with tibolone or alternatively, continuous combined oestrogen and progestogen. The oestrogen may be administered by any of the available routes. Oral, low dose progestogen therapy is given daily, the dose of progestogen being adjusted until the woman becomes amenorrhoeic.

(c) Premenstrual symptoms

These are always due to the use of progestogen and may be relieved by a reduction in dosage or change in the type of progestogen. Different women respond differently to different progestogens. The choice of the most appropriate type of progestogen for any particular patient is simply a matter of 'trial and error'.

13.5.8   Other systemic side-effects

Unfortunately, many women will experience side-effects from the systemic progestogen. A progestogen containing intrauterine contraceptive device is now available. The use of such a device in combination with oestrogen replacement therapy will have the advantages of enabling the progestogen to have a direct local action on the endometrium without significant systemic absorption, thus minimizing systemic side-effects.

13.5.9   The use of hormone replacement therapy in association with specific medical conditions

1.   Oestrogen-dependent malignancies
2.   Previous thromboembolic disease
3.   Diabetes mellitus
4.   Cardiovascular disease
5.   Abnormal smears/Cervical cancer
6.   Ovarian carcinoma
7.   Smoking

(a)   Oestrogen dependent malignancies

Breast carcinoma   Certain breast carcinomas are known to be oestrogen dependent and oestrogen replacement therapy is thus contraindicated. If such women have debilitating vasomotor symptoms they may respond to progestogen therapy which may also help to prevent bone loss.

Endometrial carcinoma   In Stage I disease with complete surgical resection and no recurrence at five years, there is no contraindication to the use of oestrogen replacement therapy. As with breast cancer, in patients with advanced disease and debilitating vasomotor symptoms, progestogen therapy may well provide symptom relief.

(b)   Previous thromboembolic disease

The low doses of natural oestrogen used in HRT have been shown to be much less thrombogenic than the synthetic oestrogens used for contraception. In deciding about the advisability or otherwise of offering HRT to a patient with a history of a previous thromboembolic episode it is necessary to distinguish between those patients in whom there is an obvious non-recurring precipitating factor such as pelvic surgery, fracture of the lower limb or prolonged bed rest and those in whom there was no obvious precipitating factor or in whom the thromboembolic event was associated with pregnancy or the use of the oral contraceptive pill.

In those patients with a non-recurring, non-hormonally related condition, HRT can be safely given without any preliminary investigation. In the remaining patients screening should be undertaken to exclude conditions such as antithrombin III or protein C deficiency.

In patients with a proven history of thromboembolic disease, transdermal patches or implants are favoured as these pass directly into the systemic circulation, thus minimizing the effect on the synthesis of factors involved in coagulation or fibrinolysis.

(c)   Diabetes mellitus
HRT has not been shown to adversely affect glycaemic control.

(d)   Cardiovascular disease
As noted previously, HRT has been shown to reduce the risk of recurrent MI in patients with ischaemic heart disease.

(e)   Previous abnormal cervical cytology or carcinoma of cervix
Neither of these conditions are oestrogen dependent and there is no contraindication to the use of HRT in women who have been treated for either of these conditions.

(f)   Ovarian carcinoma
This is not an oestrogen-dependent malignancy and there is no contraindication to the subsequent use of oestrogen replacement therapy.

(g)   Smoking
To date, there is no evidence that the use of HRT in smokers adds to the risk of cardiovascular disease and indeed, it appears to offer similar cardioprotective effects as have been demonstrated in non-smokers.

Learning Points

The menopause will affect the majority of women.

The menopause can cause quite severe symptoms and result in long-term problems, the most significant being an increasing risk of cardiovascular disease and osteoporosis.

The menopause can cause both physical and psychological problems.

Some of the problems associated with gradually diminishing oestrogen levels (the climacteric) can and do occur prior to the cessation of menstruation (the menopause)

Hormone replacement therapy is a safe and effective method of controlling both the symptoms of the menopause and the long-term and widespread pathological changes associated with oestrogen deprivation.

In women who retain an intact uterus, oestrogen should not be given 'unopposed'. At least 12 days progestogen therapy should be given each calendar month to prevent endometrial hyperplasia and neoplasia.

Hormone replacement therapy may be associated with a small increase in risk of developing breast cancer. Overall, the benefits probably outweigh the potential risks.

Hormone replacement therapy can be given by different routes. The dose and mode of administration should be determined on an individual basis.

Further reading

Drife, J.O. and Studd, J.W.W. (eds) (1990) *HRT and Osteoporosis*. Springer-Verlag, London.

Studd, J.W.W. and Whitehead, M.I. (eds) (1988) *The Menopause*. Blackwell Scientific, Oxford.

Whitehead, M. and Godfree, V. (1992) *Hormone Replacement Therapy – Your Questions Answered*. Churchill Livingstone, Edinburgh.

Problems with sexuality

14

Tony Parsons

14.1 Background

Problems with sexuality are fundamental to the understanding of gynaecology, both as the causes and the consequences of gynaecological problems. Before the clinician can help women appropriately, he or she will need to become comfortable with the subject matter. There are a number of prerequisites for this. The first is to acknowledge the underlying and integral importance of sexuality in gynaecological complaints. Despite the fact that the focus of gynaecology is the female genital tract, the underlying sexuality is often ignored or denied. A good basic knowledge of human sexual functioning is essential, as are good general interviewing and counselling skills. Finally it is necessary to be able to identify psychosexual disorders and be aware of the pelvic pathology that can be associated with them. Armed with these skills, clinicians should be prepared to ask sexual questions routinely during history taking, rather than waiting for patients to broach the subject.

14.2 Sexual dysfunctions

Masters and Johnson's pioneering work on the human sexual response has resulted in much emphasis being placed on sexual dysfunctions, i.e. an inability for one or both partners to achieve a physiological pattern of sexual response. In women this may mean problems with arousal, or inability to reach orgasm. However, a physiological response does not necessarily equate with a satisfactory and rewarding sexual relationship. The term sexual difficulties is used to describe a variety of problems which may still occur in the presence of normal arousal and orgasm, and these tend to reflect the individual's subjective feelings. Dissatisfaction with the sexual relationship may also be expressed even when there is no clearly defined problem, and this may sometimes reflect more deep-seated feelings about the relationship in general. Sexual 'deviations' or 'variations' are forms of sexual behaviour or methods of sexual arousal which fall well outside the limits of what most people would consider normal. This type of problem needs to be dealt with by people with the expertise to understand the underlying mechanisms, and will not be considered further here.

14.3 Scale of the problem

The true incidence of sexual problems in the community is unknown, and studies in this area all tend to have problems with selection biases. The Kinsey reports on male (1948) and female (1953) sexual behaviour are still the most comprehensive sources of information. However, there are a number of more recent studies which all suggest appreciable levels of problems. Frenken's (1976) study is fairly typical, but is more representative of the general population than some. This suggested 12% of men and 9% of women clearly avoiding sexual activity with a further 14% of men and 33% of women showing 'weak avoidance'. In this study 26% of men and 43% of women had some problems with arousal and enjoyment. About 33% indicated either difficulty or dissatisfaction with orgasm, and a further 5% were anorgasmic. Another study by Garde and Lunde (1980) was based on 40-year-old Danish women, and found 35% reporting some sexual problems. Few studies give good descriptions of clinic populations but overall, whereas men tend to complain of problems with their genital responses, women complain predominantly of lack of interest or enjoyment.

14.4 The role of the gynæcologist

In the light of these background levels, it is clear that the role of the gynaecologist is not to assume that every woman who has problems with her sexuality is a 'case for treatment', but rather to be able to elicit and openly discuss such problems in the context of her presenting condition. Routine enquiry about sexuality need not take up a large part of routine history taking. Kolodny *et al.* (1979) suggested that the following questions would identify 95% of women with sexual problems in clinical practice:

1. Are you currently active sexually? If so, what's the approximate frequency?
2. Are you satisfied with your sex life? If not, why not?
3. Do you have any difficulties with arousal?
4. Do you ever experience pain during intercourse?
5. Do you have difficulty becoming orgasmic?

14.5   Changes in sexuality

Sexuality is not fixed throughout an individual's life, but changes during pregnancy, following childbirth and with ageing. It may also be moulded by experience with partners, both past and present. In general, women who have rewarding sexual relationships with good levels of emotional communication with their partners will tend to adapt relatively easily to physiological changes, whereas those whose sexual relationship is less secure will be more likely to develop overt dissatisfaction or dysfunction at these times.

14.5.1   Loss of libido

It is important to discover what underlies a complaint of having 'gone off sex' or of having a loss of sexual drive. One of our prime difficulties in dealing with loss of libido is our lack of understanding about the 'desire phase' of sexual response which is perhaps the least studied and most complex part.

Animal experiments have located a number of sensors within the brain which control sexual desire, similar to those controlling sleep, appetite, etc. These are located in the limbic system, the archaic part of the brain which controls those primitive desires which are related to survival. Other sensors have also been described in the hypothalamus and pre-optic region. Both activating and inhibiting centres have been demonstrated and hormones have a modulating effect on both. The relative roles of oestrogen and testosterone are still controversial, but testosterone certainly seems to be an important influence in both sexes. Loss of testosterone tends to affect desire before having a direct effect on sexual response, and this may be clinically relevant, especially following castration or natural menopause. The centres controlling sexual drive have widespread neuroanatomical connections with the cerebral cortex, so an individual's life experiences and the values of the society in which she is reared will be important determinants of her libido.

A careful history is important in women complaining of loss of libido. Although some will have had normal levels of sexual desire which have recently deteriorated, there will be others for whom an unsatisfactory situation has merely worsened. Further details of a woman's previous sexual experiences may be important at this stage. It is not unusual to see women complaining of a loss of libido when the true problem may be one of a primary lack of drive. Early sexual experimentation, particularly in the context where sex is 'forbidden' may have provided initial levels of interest and excitement, but it may not have been possible to translate this into interest in an ongoing sexual relationship.

Development of sexual drive in women starts at puberty and tends to gradually increase, peaking at around the age of 40. Adverse factors in a woman's early life may result in delaying of psychosexual development, and this commonly presents as 'general sexual dysfunction' where there is low sexual drive and often positive avoidance of sexual contact as well as an inability to become aroused.

Libido in women is probably rather more easily suppressed than in men, and there are a number of factors which may be responsible for this. One of the commonest of these is clinical depression. Libido is often one of the first things to go in a depression and one of the last to come back after recovery. This is particularly important because a relatively large proportion of women seen in gynaecology outpatient clinics are clinically depressed.

The ability to be interested in and enjoy sexual experiences is usually dependent on an ability to relax and to clear one's mind of other issues. Stress and worries, which tend to inhibit relaxation and give rise to intrusive distractive thoughts, are likely to impair sexual response. Repeated lack of enjoyment of sexual experiences can then be reflected in a lower sexual drive or even an 'aversive' response where the individual avoids sexual contact and may become tense and anxious even when her partner expresses physical affection in a non-sexual way. A woman may, for example, find herself quite unable to tolerate her partner kissing her or putting his arms around her.

14.5.2 Factors affecting libido

Many drugs have an effect on libido, and a careful history should be taken of both social and prescribed drugs. In general, small amounts of sedatives or anxiolytics, including alcohol, may increase sexual desire, particularly where libido is being reduced by general sexual inhibition or by anxiety. Amphetamines and cocaine are said to enhance sexual desire at low doses, whereas most other drugs of abuse have either no effect or an inhibitory one. Antipsychotic agents generally tend to decrease libido, as may lithium, but antidepressants generally have no effect. The effects of other drugs are unpredictable and medication should always be thought of as a possible explanation for a secondary lack of libido.

Exogenous sex steroids might be expected to have a clear-cut effect, but in fact the situation is more complicated. With the combined pill, other factors may override the hormonal effect, so that increased desire may be due to a decreased fear of pregnancy, whereas decreased desire may be due to changes in the relationship or to a lack of excitement in the absence of any risk of pregnancy. The role of hormone replacement therapy and of treatment with testosterone will be discussed later in this chapter.

Endogenous sex steroids clearly do play a role in libido, and so it is not surprising that there are certain times in women's lives when they are more prone to a loss of sexual desire, though these situations are usually multifactorial. Low libido is common soon after childbirth, though clearly the changed hormones are not the only factor. The woman's attention is often taken up with her new baby. She may also be very tired, and after a first child worries about how her body has changed may have an inhibitory effect.

Small fluctuations in mood and well-being with the menstrual cycle are common. Many women will find that their libido tends to increase during or following menstruation, while the oestrogen levels are rising, and to be at a minimum premenstrually. This of course will be aggravated by any other premenstrual symptoms such as tension, irritability and bloating.

Changes in libido at the menopause are unpredictable. Some women actually find their sex drive increases, perhaps partly because of relief from menstrual symptoms and from fear of pregnancy. Others experience a deterioration and in some this may be secondary to loss of sexual enjoyment because of vaginal dryness and reduced vulval and vaginal sensitivity, consequent upon falling oestrogen levels. As with the other examples previously mentioned, this change in libido is often related to a combination of factors. Many women experience lowered mood as they go through the menopause and the climacteric happens at a time of high levels of other stresses and domestic upheaval for many women.

Oestrogen replacement will restore libido in many women, though this should be backed up by some counselling of both partners about the changes in sexuality which accompany the menopause. Subcutaneous testosterone implants can be very helpful for those who do not respond fully to oestrogen replacement, particularly where the pattern is of loss of interest prior to a loss of response. This is a normal pattern which one sees after bilateral oophorectomy, where the ability to undergo a physiological sexual response remains present despite the absence of any interest.

Generalized debilitating illness will usually have a negative impact on libido, particularly if a woman is anxious about the nature and consequences of the illness. This is even more true in gynaecology where many women have quite disturbing mental pictures of the consequences of what clinicians may regard as routine operations such as vaginal repairs and hysterectomies. Gynaecological surgery does, of course, have an impact on sexual functioning, and it is important that women should be carefully counselled about this.

14.6   Physical problems preventing intercourse

Physical problems preventing intercourse	
Male Problems	Failure to achieve an erection
Vaginismus	Secondary and primary
Absent vagina	
Imperforate hymen	
Chronic painful conditions	Episiotomy scar, endometriosis

As many as 5% of all couples will not consummate a relationship within two years of marrying or living together. This can often be because of male problems, but probably the commonest female problem explaining non-consummation is vaginismus. This is the development of a reflex contraction of the pubococcygeus muscle on attempts at penetration, which results in complete closure of the vaginal orifice. The tightened muscle then feels tender and further attempts at penetration cause pain and further spasm of the muscles. Women with vaginismus will often present believing that there is an actual blockage in the vagina, describing the sensation that it feels as if their partner is pushing against a bone. Libido in women with vaginismus is very often completely normal as is the ability to achieve an orgasm.

Primary vaginismus (i.e. vaginismus which is present from first attempts at intercourse onwards) is usually associated with a history of an inability to use tampons during menstruation. Fantasies about the nature of the vagina may play a role in the original causation, as may fear of pain or damage on penetration. However, by the time women present with vaginismus, the main problem is usually that the reflex contraction of the pubococcygeus muscle has been well established. Clearly, forcible examination at this stage will only worsen the situation. It probably is appropriate to ask the woman whether she feels able to take up the usual position for examination, as it is sometimes possible to confirm the absence of any obstruction without causing distress. Furthermore, in many women the reflex will have generalized further so that they will be unable to take up this or even to separate their knees. If it is not possible to carry out a physical examination, this should usually be deferred until a behavioural approach to the problem has taught the woman how to overcome the reflex and take back control of the pubococcygeus muscle.

Some women, of course, do have an actual obstruction to penetration and these should not be missed. This would include women with an absent vagina (where the history of primary amenorrhoea should have been elicited) and women with a rigid, thick hymen. Secondary vaginismus (i.e. vaginismus occurring in the woman who was previously able to enjoy full penetration without any problems) should be viewed differently. Although this may be related to powerful feelings of fear or anger engendered by the relationship, it is more likely to be secondary to a painful lesion within the vagina. This can particularly occur where there is abnormal tenderness in an episiotomy scar.

14.7   The impact of disease on sexual function

Organic disease may impact on sexual function in a number of ways. Debilitating systemic illness will tend to reduce levels of sexual interest. Fears about an illness and its impact on both health and attractiveness may reduce interest and the ability to respond. Gynaecological conditions will, of course, often have an impact on sexuality, as may its treatment. This particularly applies to conditions and surgery affecting the vulva or vagina. Dyspareunia (painful intercourse) can be a consequence of vulval or vaginal infections or of scarring from previous surgery. Sexually transmitted infections and pelvic inflammatory disease may cause discomfort, but the ability to continue to enjoy a sexual relationship may also be hampered by feelings of shame or of anger directed towards the partner.

It has long been controversial whether a hysterectomy adds to the development of psychosexual problems, but it does seem that for the vast majority of women for whom hysterectomy is indicated, sexual relationships subsequently improve. The uterus is not a necessary component for sexual enjoyment, though there is considerable anecdotal evidence to suggest that some women gain important erotic sensation from pressure against the cervix or uterus during intercourse. Some are also more aware of uterine contractions as a component of their orgasm. Gynaecological cancers may cause sexual disability in several ways. Since libido, particularly in women, is often closely linked to mood, it can be suppressed by even mild levels of depression, which may occur as part of the illness process or as a reaction to the diagnosis. Cancer surgery can often be mutilating and may radically modify an individual's body image. Studies of women who have undergone radical surgery for gynaecological cancers have shown significant deterioration both in their body image and in their subsequent sexual relationships. Radiotherapy, particularly for cervical cancer, may produce a degree of vaginal stenosis with reduced lubrication and vaginal soreness, and for this reason it is not usually the treatment of choice in the younger women with an early stage cervical cancer.

14.8   Psychosexual counselling

Psychosexual counselling is the term used to describe the way in which we try to help people who are experiencing sexual difficulties or dysfunction. There are a number of different approaches to this which range from the highly behavioural to the psychodynamic.

Fortunately, the majority of women who tell us about their sexual problems do not require this level of intensive specialist help. Annon (1976) devised a model by means of which the appropriate level of response can be selected for each problem presented to the clinician.

This model, also known by the acronym PLISSIT, refers to four levels of response:

P	Permission
LI	Limited information
SS	Specific suggestions and
IT	Intensive therapy

With each step through the PLISSIT model the number of remaining problems can be reduced. 'P' stands for permission. Patients may tell us about aspects of their sexuality in order to share their anxieties and be reassured that the present situation can be allowed to continue. Soon after childbirth, when a woman's attentions are taken by her baby and she may be getting little sleep, her partner may nonetheless be wanting to restore the level of sexual activity which they enjoyed before the pregnancy. She may need 'permission' to be tired and less interested in sex. A woman who enjoys her sexual experiences, but is anorgasmic, may not necessarily be asking for therapy. She may be wanting to hear that this is not unusual and that however much her partner may want her to be orgasmic, her own sexual response is her own property.

'LI' stands for limited information. For example, women undergoing hysterectomy may understandably be anxious about the effect of this on their future sex life. A simple explanation about any hormonal changes, a diagram to show how the vagina will be converted into a cul-de-sac, but its overall dimensions unchanged, and some discussion of the avoidance of deep penetration until the vaginal vault regains its elasticity, may be all that is needed. Many women are unclear about the role that their womb plays in their sexual response, and again simple information about this may reduce their anxieties.

'SS' stands for specific suggestions. During pregnancy, women often experience changes varying from alterations in sexual desire to changes in sensitivity and preference of their erogenous zones. Simple suggestions about relearning about their body's response and advice about comfortable and enjoyable positions for intercourse in late pregnancy may be all that are required.

'IT' stands for intensive therapy, and this is the more detailed psychosexual counselling which some women will need. Although in principle this should be within the remit of any gynaecologist, the limitations of the British healthcare system make this impracticable for most. To carry out psychosexual counselling one needs, in addition to the appropriate sensitivity and training, the possibility of seeing the same couple on repeated occasions over a relatively short period of time and in a setting which is free of interruptions and conducive to relaxation and openness. This is not possible in a typical gynaecology outpatient clinic.

The most common categories of sexual dysfunction presenting among women are:

1.   Dyspareunia
2.   Vaginismus
3.   Orgasmic dysfunction
4.   General sexual dysfunction
5.   Low or loss of libido.

These are the types of problems for which psychosexual counselling may be appropriate. Problems of low libido may also be treated in this way, though the outcome tends to be less satisfactory.

Therapy starts with a detailed assessment to gain an understanding of the presenting problem and to elucidate any underlying organic factors. A sexual problem history should be taken with the emphasis on the current problem. Specific questioning will be needed to gain a full description of the presenting complaint. Focusing on this rather than going straight to taking a more general sexual history has a number of advantages. First, it reassures the patient that you are taking her problem seriously. Second, it often provides the necessary information for recognizing possible organic components.

When a clear picture of the problem as it exists has been obtained, a history of this problem should then be obtained. This will include some information about the first time the problem occurred, and of how it has developed. It is important to know whether a sexual dysfunction is situational (e.g. if it occurs with one partner and not another). Obvious aggravating and ameliorating factors should also be asked about and in the case of pain, the standard eight questions which are asked about any complaint of pain should be included. Finally it is worth trying to get some idea of the woman's own views of the cause of the problem and her partner's reaction to it. Attempts are often made to distinguish between sexual and 'marital' problems, but unless the relationship is really unsatisfactory, there seems to be little advantage to this approach.

When a full history and physical examination has been carried out, it is usual to set a structure for the counselling process. Although different practitioners will have different approaches to psychosexual counselling, the most popular and, many would say, the most effective has come to be described as sex therapy. Sex therapy is a technique of combining prescribed sexual experiences (a behavioural approach) with psychotherapy. Typically this approach is both client-centred and couple-centred. Any attempt at intercourse is usually 'banned' at this stage and physical and sexual experiences are prescribed in very small steps. The woman's or couple's experience in trying out these steps can then be examined at a subsequent visit. Certain common factors are involved in the maintenance of chronic sexual difficulties, and these often need to be tackled separately. They include a lack of basic knowledge about sexuality, a failure to communicate (especially on matters with a sexual or emotional content), goal orientation, sexual anxiety and spectatoring. All these factors have been well-described by Masters and Johnson and Kaplan, and are usually present in those with continuing sexual difficulties. Repeated sexual disappointment leads to further anxiety at each encounter, and this in turn leads individuals to watch their body's responses rather than to allow arousal to take place spontaneously. It is worth noting that couples who previously enjoyed a varied and imaginative sex life may often restrict the expressions of their sexuality further and further when a problem arises until their behaviour becomes highly stereotyped and geared only to achieving a specific sexual goal which their problem has defined.

This type of sex therapy as described is normally used as a relatively brief intervention for couples with sexual problems. Typically treatment sessions might be held at fortnightly intervals with a standard initial contract of up to twelve sessions. Success rates vary widely. Approximately 60% of women with a sexual dysfunction derive significant benefit from sex therapy. It is hard to be more precise about success rates because of the difficulty in defining success. Often women who do not achieve the prime goal of therapy, such as becoming orgasmic, may still find their sexual response and the quality of their sexual relationship enhanced considerably. Vaginismus is highly responsive to a behavioural approach, and successful relaxation of the pubococcygeus muscle can be achieved in over 90% of cases. However, it is not unusual to then find problems (usually difficulty in maintaining an erection) developing in the male partner, and so the actual consummation rate is rather lower. At the other end of the scale, problems of libido, particularly primary lack of libido, tend to be substantially more difficult to treat. There is little information about subsequent relapse rates after treatment, though it is clear that this occurs. Although Masters and Johnson reported only a 5% relapse rate in their subjects, this figure would probably be optimistically low for most therapists.

14.9   Rape counselling

Rape is a crisis of immense proportions. It is often life-threatening and the rape victim may have experienced aggression, humiliation and emotional trauma, as well as pain and physical trauma. It is often characterized as a crime which degrades, dehumanizes and violates the victim's sense of self.

Rape changes the way the victim feels about herself, others and the world around her. Rape counselling starts with the first contact with the victim. When first reported, it is normal for the woman to be examined either by a police doctor or by a gynaecologist experienced in this field. This requires extreme sensitivity and a sense of proportion needs to be kept between the need to investigate the crime and the need to help the victim.

In the initial phase following a rape, a woman typically may experience gross anxiety, a state akin to shock, and a sense of disbelief. This may last from days to weeks. Crisis intervention techniques are needed in this context. The woman may need someone to remain with her for some time, usually in her home, until she starts to move to a less disorganized emotional state. Once the initial shock phase begins to pass, rape counselling can take place. In many ways this is similar to bereavement counselling. A typical sequence of reactions of denial, anger and guilt will follow. It is not unusual in the denial phase for women to present themselves to the outside world as if they feel that they have 'got over' the trauma. The task of counselling is to provide the victim with an empathic person with whom she may share her feelings, and work through these reactions. As well as encouraging the woman to share and acknowledge her emotional experience, it may be necessary to repeatedly challenge her tendencies to take the blame for what has happened to her. Adequate rape counselling at the time of the trauma may help to minimize the long-term consequences, but continuing sexual difficulties in this group are not uncommon and may need dealing with in their own right.

14.10   Sexuality in elderly women

In our ageist society, it is not unusual for doctors (and their patients) to believe that sex is the prerogative of the young and healthy. In fact, although sexual interest does seem to decline more rapidly in women than in men after the age of 50, studies have found about 1 in 6 of 70-year-old women are sexually active. It must be remembered that with increasing age, an increasing proportion of women do not have partners, and it is more difficult for an older woman to find a new partner than it might be for a man.

As men and women age there is a tendency for the speed of sexual response to slow. Vaginal lubrication and sexual arousal is highly oestrogen dependent, and can be shown to be reduced for some years before the menopause. However, with increasing duration of oestrogen deficiency, lubrication is reduced still further. There is some controversy about whether continuing sexual activity modifies this process. However, there is also a reduction in elasticity of the tissue of the vaginal wall, and surrounding connective tissue (because of loss of collagen). Orgasmic ability does not seem to change, though the uterine contractions of orgasm may become painful in some women. Male partners also become increasingly likely to develop difficulties, and this is probably the main factor in the reduction of coital frequency in older couples. In the presence of these physiological ageing changes, couples with previously satisfactory sexual relationships can maintain an enjoyable if less active sex life. It is important that gynaecologists should not make this more difficult by failing to take the older woman's sexual needs seriously. In addition to counselling the elderly couple about the natural physiological changes, it may be important to provide help to maintain good oestrogen levels in the vagina. Vaginal surgery, particularly posterior vaginal wall repairs, may aggravate the natural narrowing and loss of elasticity of the vagina, and should be carried out with considerable care in women who are still sexually active or who are at all likely to wish to resume sexual activity.

Learning Points

Problems with sexuality can both cause and be caused by gynæcological problems.

Most women are reluctant, initially, to discuss sexual problems. The clinician should be prepared to ask sexual questions.

Pregnancy, childbirth, the menopause and ageing can all affect sexual function.

Women in stable emotional relationships adapt more readily to factors that may influence sexual function, i.e. the menopause.

Organic disease and/or congenital abnormalities may cause sexual difficulties by either preventing satisfactory intercourse or by affecting arousal.

Gynæcological interventions especially vulval and vaginal surgery, can influence sexual function.

Loss of libido associated with the menopause may respond to oestrogens with or without additional testosterone.

Sexual dysfunction may also be a problem in elderly women.

Further reading

Annon, J. (1976) *Behavioural Treatment of Sexual Problems: Brief Therapy*. Harper and Row, Harperstown, MD.

Bancroft, J. (1989) *Human Sexuality and its Problems*. Churchill Livingstone, Edinburgh.

Bing, E. and Colman, L. (1977) *Making Love during Pregnancy*. Bantam, New York.

Clark, M.E. and Magrina, J. (1983) *Sexual Adjustment to Cancer Surgery in the Vaginal Area*. University of Kansas, Kansas.

Comfort, A. (1978) *Sexual Consequences of Disability*. George F. Stickley, Philadelphia.

Frenken, J. (1976) *Afkeer van seksualiteit*. Van Loghum Slaterus, Deventer.

Garde, K. and Lunde, L. (1980) Female sexual behaviour: a study in a random sample of 40-year-old women. *Maturitas*, **2**, 225.

Kaplan, H.S. (1979) *Disorders of Desire*. Brunner/Mazel, New York.

Kinsey, A.C., Pomeroy, W.B. and Martin, C.E. (1948) *Sexual Behaviour in the Human Male*. W.B. Saunders, Philadelphia.

Kinsey, A.C., Pomeroy, W.B., Martin, C.E. and Gebhard, P.II. (1953) *Sexual Behaviour in the Human Female*. W.B. Saunders, Philadelphia.

Kolodny, R.C., Masters, W.H. and Johnson, V.E. (1979) *Textbook of Sexual Medicine*. Little Brown, Boston.

Masters, W.H. and Johnson, V.E. (1966) *Human Sexual Response*. Little Brown, Boston.

Valins, L. (1988) *Vaginismus*. Ashgrove Press, Bath.

15 Short notes on other conditions

David Luesley

15.1 Congenital abnormalities of the female genital tract

The female genital tract forms from the Mullerian system and the genital sinus. The latter forms the vulva and lower part of the vagina whereas the former forms the uterus and tubes. The Wolffian system which in a male fetus would form the internal genitalia regresses although it is not uncommon to find residual remnants of this system in the female. These remnants form Hydatids of Morgagni (small fimbrial fluid-filled cysts) and also Cysts of Gartner's ducts (see below). The latter can form cysts in the lateral wall of the vagina.

Abnormalities range from total absence of structures to minor degrees of midline fusion problems.

15.1.1 Imperforate hymen

The hymen may completely occlude the vagina leading to retention of menstrual blood at the time of puberty. The collection of blood in the vagina behind the hymen is called a haematocolpos and if the uterine cavity becomes distended then this is termed a haematometra. Women suffering from these conditions will classically give a history of amenorrhoea but cyclical period-like pains. Development is otherwise normal. If the hymen is perforate but with only small perforations, menstruation will be normal but women may complain of sexual difficulties and difficulty with the insertion of tampons.

15.1.2 Uterine fusion abnormalities

The uterus may be completely duplicated along with the cervix and upper vagina if the two Mullerian Ducts fail to fuse. This will lead to a condition of bicornuate uterus and a double cervix (Bicollis) and a vaginal septum. These conditions can result in excessive menstruation, problems with miscarriage and later in pregnancy, malpresentation. A vaginal septum can cause problems in labour and may also cause dyspareunia.

Lesser degrees of fusion abnormality lead to Bicornis unicollis (double uterus but one cervix) (Figure 15.1), septate uterus and arcuate uterus. Occasionally both or one of the Mullerian ducts totally fails to develop leading to absent uterus or a unicornuate uterus.

The not infrequent association of renal tract abnormalities along with fusion abnormalities should prompt thorough investigation of the renal system if uterine abnormalities are present.

If the gonads themselves fail to develop as say in Turner's syndrome (XO), the genitalia are usually hypoplastic, however Mullerian development does occur.

15.1.3   Absent vagina

If the urogenital sinus fails to develop there will be an absence of vaginal development although outwardly all else will appear normal. This is in contrast to testicular feminization where the sinus develops leading to a small pouch but there is no development of the Mullerian system. This is the result of androgen insensitivity. The Mullerian system is suppressed (by Mullerian inhibiting factor produced by the

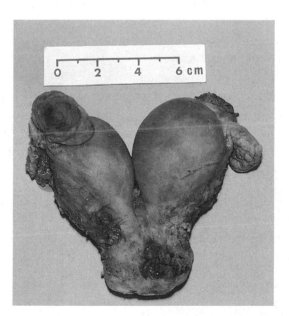

Figure 15.1 Bicornuate uterus with one cervix (unicollis).

testis), but the development of the Wolffian system is not stimulated and apart from the internal genitalia, development is as a female.

15.1.4   Clitoral hypertrophy

In a normal female where the tissues are normally androgen sensitive, raised levels of androgens, either as a result of drug exposure or to virilizing tumours will result in virilism. A virilized female infant will have signs of clitoral hypertrophy and in extreme cases may be difficult to tell apart from a male infant.

15.2   Adult virilism

Adult females exposed to testosterone or long-term exposure to weaker androgens will also develop signs of masculinization (virilism). These signs include clitoral hypertrophy, deepening of the voice, breast atrophy, hirsutism and male hair distribution.

15.2.1   Hirsutism (excess body hair)

This can occur on its own and is frequently the most concerning symptom that women complain of. The majority of women who notice excess body hair are, however, not virilized. The excess hair is either an isolated phenomenon, constitutional or within normal limits (but a low threshold of complaint).

Hirsutism should prompt an investigation of serum androgens and if raised this could either reflect a virilizing tumour such as an androblastoma or polycystic ovarian syndrome. The latter is characterized by oligomenorrhoea, hirsutism and a tendency to obesity. The ovaries are smooth, slightly enlarged and on section show multiple small cysts. Ovulation is unusual in these instances and the ovary produces continuous amounts of oestrogen and several weaker androgens but no progesterone.

15.3   Cervical ectropion (Figure 15.2)

Two types of epithelium converge at the anatomical external os. Endocervical glandular epithelium lines the endocervical canal, is one

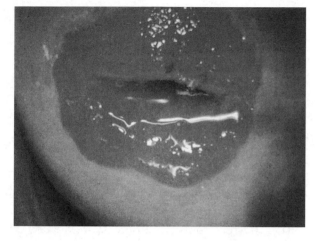

Figure 15.2 Cervical ectropion.

cell thick and produces mucus. Squamous epithelium lines the ectocervix, is stratified in many layers and is continuous with the vaginal squamous epithelium. This cervical squamous epithelium is known as native squamous epithelium. A second type of squamous epithelium exists on the cervix and this results from squamous metaplasia of glandular epithelium under the influence of the low vaginal pH. The area between native squamous and glandular epithelium is composed of this type of metaplastic epithelium.

A cervical ectopy (ectropion) often erroneously called an erosion, is a situation where the junction between squamous and glandular epithelium is sited on the ectocervix. Various degrees of extension may be seen from situations where there is only a small circumferential extension beyond the anatomical external os to situations where nearly the whole ectocervix is covered by glandular epithelium.

Ectropion is seen more frequently in pregnancy and in women using combined oral contraception. It is hardly ever seen in postmenopausal women.

Ectropion usually presents as a result of a heavy mucoid vaginal discharge (glandular epithelium produces mucus) or as a result of postcoital bleeding (the single cell thickness renders this epithelium easily traumatized). The presence of an ectropion does not predispose to malignant transformation.

Treatment, which is only indicated if associated symptoms are present, is by destruction of the ectocervically placed glandular epithelium by either cryocautery or diathermy or laser.

15.4 Cervical polyps

These are relatively common and present either in association with abnormal vaginal bleeding (postcoital or intermenstrual or postmenopausal) or are chance findings at a speculum examination usually performed at the time of taking a cervical smear. They occur as a result of polypoid overgrowth of glandular epithelium.

They are treated by simple avulsion if on a slim pedicle but broad based polyps may require avulsion with suturing of the base under general anæsthesia. Very occasionally a submucous fibroid may prolapse through the cervix and present as a cervical polyp. These are treated by either hysterectomy (if appropriate) or by transcervical myomectomy.

When associated with abnormal uterine bleeding, particularly postmenopausal bleeding, curettage of the endometrium should also be performed.

Cervical polyps are rarely malignant although histological examination of all avulsed material should always be performed.

15.5 Nabothian follicles

These are mucus retention cysts occurring as a result of squamous metaplasia. This process occasionally leads to the entrapment of mucus-producing glandular epithelium beneath a squamous surface. The condition is benign, rarely if ever produces symptoms and does not require treatment.

15.6 Vaginal adenosis

In utero exposure of female fetuses to diethylstilboestrol leads to abnormal development of the cervix and transformation zone. The whole cervix and upper third of the vagina may be covered in glandular epithelium (vaginal adenosis), the cervix may appear to be hooded and have a cocks comb appearance. Few cases of this complication are seen now since the use of diethylstilboestrol in pregnancy (in an attempt to prevent spontaneous abortion) has long since been discontinued. These women, however, remain at an increased risk of developing adenocarcinoma of the vagina.

15.7   Cervical stenosis

This almost always occurs as a result of cervical surgery and this is usually a cone biopsy performed as part of the management of patients with abnormal smears. Varying degrees of stenosis can occur from mild asymptomatic narrowing of the external os to total occlusion of the canal (Figure 15.3).

Dysmenorrhoea and even amenorrhoea with associated haematometra can occur necessitating dilating procedures and in some cases even hysterectomy.

Stenosis may also affect the ability to take an adequate cervical smear. In some postmenopausal women stenosis can occur as a natural phenomenon.

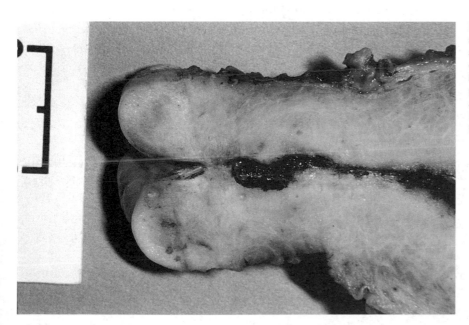

Figure 15.3 A hysterectomy specimen showing complete stenosis of the cervix. This occurred after a large knife cone biopsy had been performed. The patient became amenorrhoeic and experienced cyclical pain. Note the old blood trapped behind the stenotic segment.

15.8 Cervical incompetence

This diagnosis is usually made after a miscarriage. The classical presentation of cervical incompetence is repeated painless mid-trimester abortion.

Incompetence of the internal os usually results from a surgical procedure such as excessive cervical dilation or from a very large cone biopsy. In the former it is not unusual to find a history of previous dilation at the time of termination of pregnancy. Sometimes the os can be seen to be wider than normal on a hysterogram but in others there are no obvious signs and management is based purely on the history.

Management takes the form of insertion of a support suture (Shirodkhar or McDonald's suture). This is a purse string type of suture inserted around the cervix in early pregnancy and removed prior to the onset of spontaneous labour.

15.9 Ovarian cysts

The normal ovary will form follicles on a monthly basis. These can be seen on ultrasound scanning and are not pathological. If ovulation does not occur and the follicle continues to enlarge then a follicular cyst may result. These are seldom larger than 5 cm in maximum diameter. They are invariably associated with disturbance of the menstrual cycle. Very occasionally they may result in pain or may rupture and bleed. Failure of the corpus luteum to regress may result in a corpus luteum cyst. These may also rupture and bleed and very occasionally present as an acute emergency with haemoperitoneum.

The majority of pathological ovarian cysts are benign. They include serous and mucinous cystadenomas. They may become very large, leading to abdominal distension. Torsion and haemorrhage can occur, leading to acute presentations. Treatment is either by cystectomy, which attempts to conserve ovarian tissue in the affected ovary, or by oophorectomy.

Endometriosis can also cause ovarian cysts. These tend to be full of altered blood and are called chocolate cysts. Malignant ovarian cysts are seen more frequently in older women and should be suspected if the cyst contains solid elements and or excrescences. It is always wise to

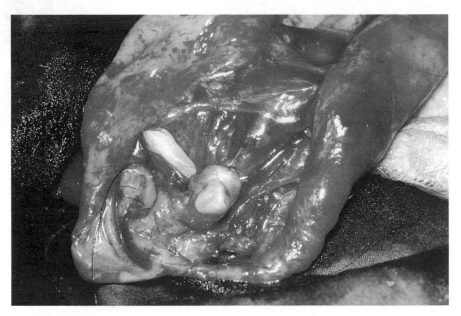

Figure 15.4 An opened ovarian dermoid cyst (benign cystic teratoma) showing teeth.

carefully inspect the outer and inner components of any cyst removed and if there is doubt as to its malignant potential multiple blind biopsies should be taken from the peritoneal surfaces, and from the remaining ovary.

In girls and young women germ cell tumours are more common. It is not possible however to determine the histological subtype just from examining the cyst although there are some changes which might raise suspicions.

Dermoid cysts are benign and contain elements of ectoderm, endoderm and mesoderm. They are usually solid and not infrequently are chance findings. They will often contain recognizable elements such as teeth (Figure 15.4) and hair, usually embedded in sebaceous material. Very rarely, malignant change can occur in one of the elements. Squamous cancer is the most likely malignancy that will occur in dermoid cysts.

15.10 Trophoblastic disease

These diseases largely arise from the placenta, i.e. they must be preceded by a conception. Choriocarcinoma can, however, arise in the ovary as a primary germ cell neoplasm.

Figure 15.5 Ultrasound appearances of a hydatidiform mole showing the classic 'snowstorm' pattern.

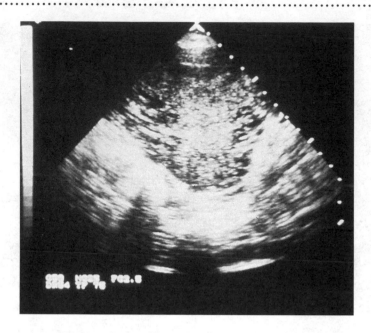

The spectrum of trophoblastic diseases includes hydatidiform mole (benign), invasive mole (locally invasive) and choriocarcinoma (frankly malignant). Hydatidiform mole is now usually diagnosed by ultrasound scan performed in early pregnancy. A classic snowstorm appearance (Figure 15.5), relating to the multiple hydatidiform cysts, is seen. All of these tumours produce βHCG and therefore will have positive pregnancy tests. The uterus may be large for dates. Treatment is by suction evacuation. Products should be sent for histological assessment. Sometimes fetal tissue can be identified, this will almost always be a triploid fetus and the triploidy seen in such pregnancies is described as a partial mole. The majority of molar pregnancies are, however, not triploid and are described as complete moles. Complete moles are thought to arise as a result of two sperm fertilizing one egg (dispermy) and the mole which is always XX derives both haploid sets from the male.

The risk of developing either invasive mole or choriocarcinoma is greater following hydatidiform mole than, for example, following a normal pregnancy or a miscarriage of a normal fetus. These patients, therefore, require heightened follow-up. This is done using serum and urinary βHCG estimations. If levels are still elevated three months after evacuation and if a further pregnancy has been excluded it is likely that persistent trophoblast is present and these patients should be considered for treatment with chemotherapy.

Choriocarcinoma is a highly malignant neoplasm with a tendency to metastasize widely. Lungs, brain, liver and bone might all be involved. They are, however, very chemosensitive tumours and even with adverse risk factors can be cured in over 90% of cases. The degree of risk is calculated from many variables that include the number and site of metastases, the serum βHCG level, the length of time between the preceding pregnancy and diagnosis and the outcome of the prior pregnancy. Patients can be classified into low, medium or high risk and the type of chemotherapy is determined from the risk status.

15.11   Benign vulval conditions

The vulva is affected by several specific disorders of the skin. These include lichen sclerosus and squamous cell hyperplasia. As well as these conditions, other skin disorders such as eczema and psoriasis can affect the vulva.

They usually present in older women and cause intense itching (pruritus vulvae). The skin may be thickened and hyperkeratotic as a result of prolonged scratching. These changes appear as white plaques and gave rise to the term leukoplakia (Figure 15.6). This is a descriptive term only and not a disease entity. Diagnosis is by directed biopsy from the most affected area. This can usually be performed under local anaesthetic and is necessary to exclude invasive and preinvasive diseases.

Lichen sclerosus is thought to be an autoimmune disorder and as a result of the protracted inflammatory reaction in the skin leads to fibrosis, scarring and atrophy. In extreme cases the anatomy of the vulva can become totally distorted with almost complete closure of the introitus. In sexually active women dyspareunia is common.

Whether lichen sclerosus is preinvasive is a matter of some debate. The condition not infrequently is seen in vulvectomy specimens removed because of carcinoma, but this could be purely an association rather than a cause.

Figure 15.6 Hyperkeratotic vulvar skin. Previously called leukoplakia. This is a case of lichen sclerosus, a condition which can be associated with hyperkeratosis.

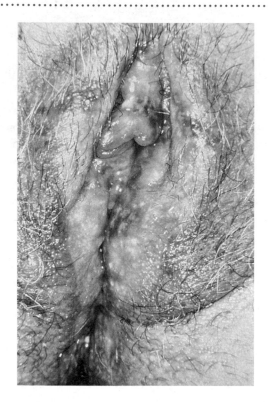

15.12   Vaginal cysts

These are either simple mucus retention cysts, cysts of Bartholin's glands or cysts forming in the remnants of Gartner's ducts. Bartholin's cysts are the most frequently recognized. They occur as a result of blockage of the ducts and can also become infected resulting in Bartholin's abscesses. They can be quite painful when infected and even painless cysts can cause problems during intercourse. They are managed by drainage and marsupialization (suturing the cyst lining to the edges of a small circumferential incision thus creating a pouch). If infected, culture of any material should be performed as gonorrhoea can cause Bartholin's abscesses.

Gartner's ducts are the remnants of the Wolffian system and are found lateral to the vagina. They are usually painless and certainly benign but because of the medial displacement caused can result in dyspareunia. They are treated by incision. It is important to perform an intravenous pyelogram prior to excision to determine whether or not there are co-existent renal tract abnormalities, particularly accessory ureters that may be closely related to Gartner's ducts.

Appendix A

Commonly used abbreviations in gynæcology

AID   Artificial insemination using donor semen. As opposed to AIH which is artificial insemination using husband's or partner's semen. The latter is usually specially prepared and injected high into the uterine cavity (HIUI or high intrauterine insemination).

AIS   Adenocarcinoma-in-situ. A glandular preinvasive condition of the cervix, less common than its squamous counterpart, CIN (vide below).

BSO   Bilateral salpingo-oophorectomy. A surgical procedure removing both ovaries and fallopian tubes.

BTB   Breakthrough bleeding. A term usually applied to the bleeding occasionally noticed by some women using combined oral contraception. The bleeding occurs while the pill is being taken and not in the pill-free week. Bleeding in the pill-free week is normal and is termed withdrawal bleeding.

CIN   Cervical intra-epithelial neoplasia. A premalignant condition of the cervix.

COC   Combined oral contraceptive. This refers to steroid oral contraceptives containing an oestrogen, usually ethinyl-oestradiol and a progestogen. POP refers to progesterone only pills. The term OCP means oral contraceptive pills and is a generic term for both combined and progesterone only preparations.

D&C   Dilatation and curettage. An operative procedure, usually performed under a general anaesthetic. The cervix is gradually dilated using a graduated set of dilators to a level where a curette can be introduced into the uterine cavity. This is then used to scrape endometrial or other tissue from the endometrium for histological analysis. The procedure is diagnostic. It is not a treatment for menstrual dysfunction.

DUB   Dysfunctional uterine bleeding. Excessive (>80 ml) or erratic menstruation where no recognizable pathology can be found. Pathologies to exclude that might account for menorrhagia or irregular cycles include fibroids, endometriosis, pelvic inflammatory disease or endometrial hyperplasia or neoplasia.

ERPOC   Evacuation of retained products of conception. Also referred to as an 'Evac'. This is an operative procedure performed under general, regional or local anaesthetic whereby residual placental and fetal tissue are scraped off the uterine walls following an incomplete abortion.

FIGO   International Federation of Obstetrics and Gynæcology. An international body that provides standards for classification such as staging procedures for gynæcological malignancies.

GIFT   Gamete intra-fallopian transfer. A method of assisted conception. Ova and sperm are inserted into the distal end of the fallopian tube under laparoscopic guidance to allow fertilization to occur in the ampullary portion of the tube.

HPV   Human papillomavirus. A group of DNA viruses. Many subtypes have been identified some of which may be oncogenic. Of particular interest in preinvasive and invasive disease of the cervix where they may be implicated in the neoplastic process (esp Types 16 and 18).

HRT   Hormone replacement therapy. Replacement of oestrogen alone or with progesterone in naturally menopausal women or in women who have had their ovaries removed. HRT is also given in situations of absent or low oestrogen production such as Turner's syndrome. In cases where the uterus remains *in situ*, i.e. not hysterectomized,

a progestogen should also be given to minimize the risk
of hyperplasia and neoplasia.

HSG   Hysterosalpingogram. An imaging technique whereby
radio-opaque dye is infused through a specially designed
applicator through the endocervical canal. Its purpose is
to highlight the endometrial cavity for abnormalities of
shape and contour and also to outline the fallopian tubes.
This will determine any blockage, the level and to some
extent the nature of the blockage. It is usually performed
as an adjunct to fertility investigations especially prior to
planned tubal surgery.

HSV   Herpes simplex virus. Two subtypes, Type I and Type II.
Both may cause genital herpes.

IMB   Intermenstrual bleeding. Any vaginal bleeding occurring
between what a woman regards as normal menses. This
may be mid-cycle as can occur at ovulation (although
usually quite light) or premenstrual spotting.

IUCD   Intra-uterine contraceptive device. Often referred to as a
coil by women. This is a plastic device, usually containing a copper component, that is inserted into the uterine
cavity as a form of contraception.

IVF   *In vitro* fertilization. Ova are collected from either natural
or more usually stimulated cycles under ultrasound guidance. The collected ova are then mixed with sperm and
fertilization and early embryogenesis supported in artificial media outside the body. The early embryo is transferred back to the endometrial cavity by the transcervical
route. This was the first type of popularized assisted
conception and is appropriate for women with either non-
functional or absent fallopian tubes where GIFT and
ZIFT would be inappropriate.

K   Koilocytosis. A short term used to describe the cytological and histological features associated with human
papillomavirus infection (see HPV).

Lap Dye   Laparoscopy and tubal patency test. An endoscopic procedure using a laparoscope to visualize the pelvic organs.
By injecting a coloured dye into the endometrial cavity

via the cervix the patency of the fallopian tubes can be assessed. Also referred to as TPT (Tubal patency test).

LAVH   Laparoscopically assisted vaginal hysterectomy.

LLETZ   Large loop excision of the transformation zone. A diathermy excision technique for the treatment of women with cervical intraepithelial neoplasia.

LMP   Last menstrual period. This refers to the first day of the woman's last period, i.e. the day that menstruation began. An average cycle lasts 28 days of which 3–7 days will be days of menstruation. The least consistent part of the cycle is the follicular phase whereas the luteal phase usually lasts about 14 days. Ovulation therefore occurs about 14 days prior to menstruation.

PCB   Post coital bleeding. Vaginal bleeding occurring either during or after penetrative intercourse. Although carcinoma of the cervix may cause this, benign conditions such as a cervical ectropion can as well. Some bleeding is not unusual following the first time a woman has intercourse.

PCT   Post coital test. Examination of a sample of cervical mucus shortly after sexual intercourse. This is done to assess the number and forward motility of sperms. It also shows if intercourse has taken place properly.

PID   Pelvic inflammatory disease. Refers to infection of the uterus, tubes or ovaries. Most likely pathogens are gonococci, chlamydia or anaerobes.

PMB   Post menopausal bleeding. This describes any episode of vaginal bleeding occurring after the menopause. The latter is usually defined as a period of 12 months' amenorrhoea which may or may not be associated with other climacteric symptoms such as hot flushes, vaginal dryness etc.

PMS/PMT Premenstrual syndrome or premenstrual tension. A cyclical disorder seen classically in the week prior to normal menstruation. Characterized by both physical (bloating, breast tenderness, oedema) and emotional (mood swings, irritability, depression and tearfulness) symptoms.

POD Pouch of Douglas. A peritoneum-lined space lying between the posterior aspect of the cervix and upper vagina and the anterior aspect of the rectum. If this prolapses down into the vagina an enterocele forms. May contain loops of small bowel.

RaFEA Radiofrequency endometrial ablation. A means of destroying the endometrium yet leaving the uterus intact. This technique employs a microwave principle to generate heat in the endometrium.

STD Sexually transmitted disease. An infection contracted by sexual contact. Also GUM, genitourinary medicine.

TAH Total abdominal hysterectomy. A surgical procedure to remove the uterus and cervix.

TCRE Transcervical resection of the endometrium. A technique where the endometrium is resected hysteroscopically using a fine diathermy loop. This is a treatment for menorrhagia that conserves the uterus.

TZ Transformation zone. An area on the cervix between squamous epithelium and glandular epithelium where dysplasia and metaplasia occur.

Vag Hyst Vaginal hysterectomy. A surgical procedure to remove the uterus and cervix but the approach is through the vagina. Usually combined with a repair procedure at the same time.

VE or PV Vaginal examination. A standard part of the basic gynaecological examination. The procedure is often called a BIMANUAL examination as the fingers of one examining hand are placed in the vagina while the other hand is used to palpate the lower abdomen.

ZIFT Zygote intrafallopian transfer. Similar to GIFT but fertilization and thus zygote formation occurs outside the body. The zygote is transferred back to the distal end of the fallopian tube to allow the early part of embryogenesis to occur in its natural environment.

Index

Note: page numbers in *italics* refer to tables, those in **bold** refer to figures.